D0131504

Geology
Rocks, Minerals, and the Earth
Expanding Science Skills Series

MVFOL

BY

LaVERNE LOGAN

CONSULTANTS: SCHYRLET CAMERON AND CAROLYN CRAIG

COPYRIGHT © 2010 Mark Twain Media, Inc.

ISBN 978-1-58037-526-9

Printing No. CD-404123

Mark Twain Media, Inc., Publishers
Distributed by Carson-Dellosa Publishing Company, LLC

Visit us at www.carsondellosa.com

HPSO 218673

Table of Contents

Introduction

Geology: Rocks, Minerals, and the Earth is one of the books in Mark Twain Media's new Expanding Science Skills Series. The easy-to-follow format of each book facilitates planning for the diverse learning styles and skill levels of middle-school students. The teacher information page provides a quick overview of the lesson to be taught. National science, mathematics, and technology standards, concepts, and science process skills are identified and listed, simplifying lesson preparation. Materials lists for Knowledge Builder activities are included where appropriate. Strategies presented in the lesson planner section provide the teacher with alternative methods of instruction: reading exercises for concept development, hands-on activities to strengthen understanding of concepts, and investigations for inquiry learning. The challenging activities in the extended-learning section provide opportunities for students who excel to expand their learning.

Geology: Rocks, Minerals, and the Earth is written for classroom teachers, parents, and students. This book can be used as a full unit of study or as individual lessons to supplement existing textbooks or curriculum programs. This book can be used as an enhancement to what is being done in the classroom or as a tutorial at home. The procedures and content background are clearly explained in the student information pages and include activities and investigations that can be completed individually or in a group setting. Materials used in the activities are commonly found at home or in the science classroom.

The Expanding Science Skills Series is designed to provide students in grades 5 through 8 and beyond with many opportunities to acquire knowledge, learn skills, explore scientific phenomena, and develop attitudes important to becoming scientifically literate. Other books in the series include Chemistry, Simple Machines, Electricity and Magnetism, Meteorology, Light and Sound, and Astronomy.

The books in this series support the No Child Left Behind (NCLB) Act. The series promotes student knowledge and understanding of science and mathematics concepts through the use of good scientific techniques. The content, activities, and investigations are designed to strengthen scientific literacy skills and are correlated to the National Science Education Standards (NSES), the National Council for Teachers of Mathematics Standards (NCTM), and the Standards for Technological Literacy (STL). Correlations to state, national, and Canadian provincial standards are available at www.carsondellosa.com.

How to Use This Book

The format of *Geology: Rocks, Minerals, and the Earth* is specifically designed to facilitate the planning and teaching of science. Our goal is to provide teachers with strategies and suggestions on how to successfully implement each lesson in the book. Units are divided into two parts: teacher information and student information.

Teacher Information Page

Each unit begins with a Teacher Information page. The purpose is to provide a snapshot of the unit. It is intended to guide the teacher through the development and implementation of the lessons in the unit of study. The Teacher Information page includes:

- National Standards: The unit is correlated with the National Science Education Standards (NSES), the National Council of Teachers of Mathematics Standards (NCTM), and the Standards for Technological Literacy (STL). Pages 61–65 contain a complete list and description of the National Standards.
- Concepts/Naïve Concepts: The relevant science concepts and the commonly held student misconceptions are listed.
- Science Process Skills: The process skills associated with the unit are explained. Pages 66–69 contain a complete list and description of the Science Process Skills.
- Lesson Planner: The components of the lesson are described: directed reading, assessment, hands-on activities, materials lists of Knowledge Builder activities, and investigation.
- Extension: This activity provides opportunities for students who excel to expand their learning.
- Real World Application: The concept being taught is related to everyday life.

Student Pages

The Student Information pages follow the Teacher Information page. The built-in flexibility of this section accommodates a diversity of learning styles and skill levels. The format allows the teacher to begin the lesson with basic concepts and vocabulary presented in reading exercises and expand to progressively more difficult hands-on activities found on the Knowledge Builder and Inquiry Investigations pages. The Student Information pages include:

1. Student Information: introduces the concepts and essential vocabulary for the lesson in a directed reading exercise.
2. Quick Check: evaluates student comprehension of the information in the directed reading exercise.
3. Knowledge Builder: strengthens student understanding of concepts with hands-on activities.
4. Inquiry Investigation: explores concepts introduced in the directed reading exercise through labs, models, and exploration activities.

Safety Tip: Adult supervision is recommended for all activities, especially those where chemicals, heat sources, electricity, or sharp or breakable objects are used. Safety goggles, gloves, hot pads, and other safety equipment should be used where appropriate.

Name: _____ Date: _____

Introductory Activity: K-W-L Rocks and Minerals

What Do We Know About Rocks and Minerals?

1. As a preliminary activity, construct a large K-W-L chart as a class. A flip chart or a large piece of freezer paper or poster board placed on a bulletin board or wall may be used. An example is shown below.

Know	Want to Know	Learned

2. Lead a class discussion to discover students' current understanding of rocks and minerals. List student beliefs under the "Know" column. Stress that this is what we believe to be the case at this point in time. Point out that this information might change as we discover more information via the activities to follow.

3. The ideas suggested by the students can provide valuable insight, which creates the basis for many instructional decisions. Students are likely to hold naïve conceptions regarding rocks and minerals; this column is where the naïve conceptions will appear. This will provide a visual representation of what students think about rocks and minerals. Be sure to list all of the ideas as written/spoken by students, despite the fact that they may be scientifically inaccurate or incomplete. As an ongoing formative assessment, return to this list as opportunities to correct naïve conceptions occur.

What Questions Do We Have About Rocks and Minerals?

1. Within the spirit of scientific inquiry, it is wise to consider student-generated questions for investigation. This can be done individually, in small groups, or as a class. It is recommended to generate a preliminary list of questions as a class. As with the section that describes what the students already know, questions may be added as the activities are completed. This mirrors the process of science, e.g., new information often yields new questions for further research.

2. Some sample questions students may have include:
    - What is the difference between rocks and minerals?
    - Are gold and silver minerals?
    - How do minerals compare with the types of minerals commonly associated with vitamins?
    - Why are minerals found in certain locations of the earth and not others?
    - What are the most common rocks and minerals around our area? Why are they found here?
    - What causes the various colors of rocks and minerals?
    - What are gems, and how do they compare to rocks and minerals? What makes a gem a gem?

Name: _____   Date: _____

Introductory Activity: K-W-L Rocks and Minerals (cont.)

3. Questions from this list may provide the basis for individual inquiry investigations.

What Have We Learned About Rocks and Minerals?

1. As conclusions are generated from the activities throughout the book, return and add them to the "Learned" column. Cross-check the "Know" column for any changes that may be appropriate. Be creative in editing the "Know" column. Draw a single line through the blatant naïve conceptions, e.g., all rocks are found on the earth's surface. Perhaps correct the naïve conception by drawing a connecting line to a better explanation found in the "Learned" column. For example, a statement in the "Know" column, "All rocks come from volcanoes," could be specified to include a description of the rock cycle found in the "Learned" column.

2. Ultimately, the goal of the book and related activities is to increase understanding of rocks and minerals by building a "Learned" column while examining the "Know" column for accuracy and depth of thought. Concurrently, new questions for inquiry that arise as investigations unfold should be considered as well.

Note: The K-W-L chart may be completed and maintained individually if the instructor wishes to track the progress of each student. Another option is to have students complete and maintain an individual K-W-L chart and also contribute to an overall class K-W-L.

Know	Want to Know	Learned
Rocks can be formed by volcanoes.	What is magma?	Magma is molten rock below the earth's surface.

Unit 1: Historical Perspective
Teacher Information

Topic: Many individuals have contributed to the traditions of the science of geology.

Standards:
   NSES Unifying Concepts and Processes, (F), (G)
   See **National Standards** section (pages 61–65) for more information on each standard.

Concepts:
- Science and technology have advanced through contributions of many different people, in different cultures, at different times in history.
- Tracing the history of science can show how difficult it was for scientific innovations to break through the accepted ideas of their time to reach the conclusions we currently take for granted.

Naïve Concepts:
- All scientists wear lab coats.
- Scientists are totally absorbed in their research, oblivious to the world around them.
- Ideas and discoveries made by scientists from other cultures and civilizations before modern times are not relevant today.

Science Process Skills:

Students will be **collecting**, **recording**, and **interpreting information** while **developing the vocabulary to communicate** the results of their reading and research. Based on their findings, students will make an **inference** that many individuals have contributed to the traditions of the science of geology.

Lesson Planner:
1. <u>Directed Reading</u>: Introduce the concepts and essential vocabulary relating to the history of the science of geology using the directed reading exercise found on the Student Information pages.
2. <u>Assessment</u>: Evaluate student comprehension of the information in the directed reading exercise using the quiz located on the Quick Check page.
3. <u>Concept Reinforcement</u>: Strengthen student understanding of concepts with the activities found on the Knowledge Builder page. **Materials Needed:** cube template, scissors, glue, pencil

Extension: Students research the history of the science of geology. Create an illustrated time line of scientists and important discoveries.

Real World Application: The first geologist to reach the moon was Dr. Harrison Hagan Schmitt. In 1972, he was part of the *Apollo XVII* moon mission with Captain Eugene A. Cernan and Commander Ronald E. Evans. He and Captain Cernan landed on the moon and spent three days on the surface, gathering scientific data.

Unit 1: Historical Perspective
Student Information

Geology is the study of the earth. Geology tries to explain how the earth was formed and how forces such as earthquakes, volcanoes, glaciers, and water change the surface of the earth. Scientists who study the earth are called **geologists**. Geologists believe the earth was formed more than 4.5 billion years ago.

The first written information we have about the earth from ancient people is a mixture of facts, superstitions, legends, guesses, and the beliefs of the time. Through study and observation, scientists were able to slowly piece together the true history of the earth.

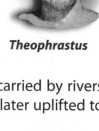

In the 300s B.C., Theophrastus (c. 372–287 B.C.) wrote a mineralogy book, *Concerning Stones*. This work gathered together for the first time all known information about rocks, minerals, and fossils.

Theophrastus

Leonardo da Vinci (1452–1519) recognized that material carried by rivers to the sea was eventually compacted into sedimentary rock and later uplifted to form mountains.

Leonardo da Vinci

In 1669, Nicolaus Steno (1638–1686), a Danish physician, discovered sedimentary rocks are laid down in a horizontal manner and layers (strata) of rock are always deposited with the oldest layers on the bottom and the youngest layers on the top. These discoveries lead to the formation of the Laws of Superposition, which scientists use to determine the order in which geolgical events took place. Nicholaus Steno's work on the formation of rock layers and the fossils they contain was crucial to the development of modern geology. The principles he stated continue to be used today by geologists and paleontologists.

Nicolaus Steno

In 1785, James Hutton (1726–1797) stated the earth was gradually changing and would continue to change in the same ways. He said these changes could be used to explain the past. Hutton, a Scottish farmer and naturalist, is known as the founder of modern geology.

James Hutton

In 1812, German scientist Friedrich Mohs (1773–1839) devised a 1–10 scale system to determine the hardness of minerals. Common objects can be used in place of the mineral on Mohs' scale to determine a mineral's hardness. By scratching unknown minerals with the suggested object in Mohs' scale, scientists can identify collected specimens.

Friedrich Mohs

In 1815, William Smith (1769–1839) was the first to use fossils to tell the age of rock layers (**strata**). He published the first geological maps showing the strata of England.

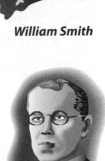

In 1830, Sir Charles Lyell (1797–1875), a British geologist, published the first volume of *Principles of Geology*. It was one of the most important events in the development of geology. He wrote that geological features take shape, erode, and reform at a constant rate through time. He was knighted for his scientific accomplishments in 1848.

William Smith

Sir Charles Lyell

In 1907, Bertram Boltwood (1870–1927), using the radioactive decay method, dated Earth's age as somewhere between 400 million and 2.2 billion years. This technique has been used since 1907, but advances in technology and knowledge of atomic structure have shown the earth to be even older.

Bertram Boltwood

In 1912, Alfred Wegener (1880–1930) proposed his **Continental Drift Theory**. He believed that the continents once formed a supercontinent called Pangaea. The continents gradually broke apart, forming the seven continents. The continents slowly drifted to their present positions.

Alfred Weneger

In 1960, Harry Hess (1906–1969) established that the surface of the earth is broken up like a jigsaw puzzle into enormous plates that move. This theory, called **Plate Tectonics**, helped support the idea that the continents drift on the earth's surface. It also explained the occurrence of mountains, volcanoes, and other geological features.

Harry Hess

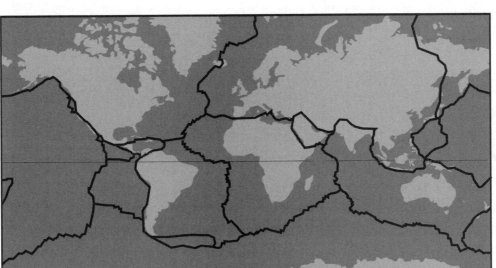

Major Plates of Earth's Crust

Name: _____   Date: _____

Quick Check

Matching

_____ 1.   geologists   a.   wrote *Concerning Stones*

_____ 2.   Alfred Wegener   b.   the study of the earth

_____ 3.   Theophrastus   c.   Continental Drift Theory

_____ 4.   Harry Hess   d.   Plate Tectonics Theory

_____ 5.   geology   e.   scientists who study the earth

Fill in the Blanks

6.   In 1907, _____ _____, using the radioactive decay method, dated Earth's age at somewhere between 400 million and 2.2 billion years.

7.   In 1812, German scientist _____ _____ devised a 1–10 scale system to determine the hardness of minerals.

8.   _____ _____ recognized that material carried by rivers to the sea was eventually compacted into sedimentary rock and later uplifted to form mountains.

9.   In 1830, Sir Charles Lyell, a British geologist, published the first volume of _____ _____ _____.

10. _____ _____ explained the occurrence of mountains, volcanoes, and other geological features with the Plate Tectonics Theory.

Multiple Choice

11. He is known as the founder of modern geology.
    a.   Friedrich Mohs   b.   James Hutton
    c.   Nicolaus Steno   d.   Leonardo da Vinci

12. His Laws of Superposition helped scientists determine the order in which geological events took place.
    a.   Theophrastus   b.   Sir Charles Lyell
    c.   Nicolaus Steno   d.   James Hutton

13. He published the first geological maps showing the strata of England.
    a.   William Smith   b.   Bertram Boltwood
    c.   Leonardo da Vinci   d.   Nicolaus Steno

Name: _____   Date: _____

Knowledge Builder

Activity: Biographical Cube

Directions: Research one of the people found on the Historical Perspective pages. Using this information, create a biographical cube.

1. Write the name of the scientist on one face of the cube. Glue a picture of the scientist on this face also.
2. Create an illustration that represents the important contribution the person made to the science of geology. Place it on one face.

3. Write other important facts, events, data, etc., on the other faces of the cube.
4. Cut out the cube and fold along the dotted lines.
5. Glue the tabs in place to form a cube.

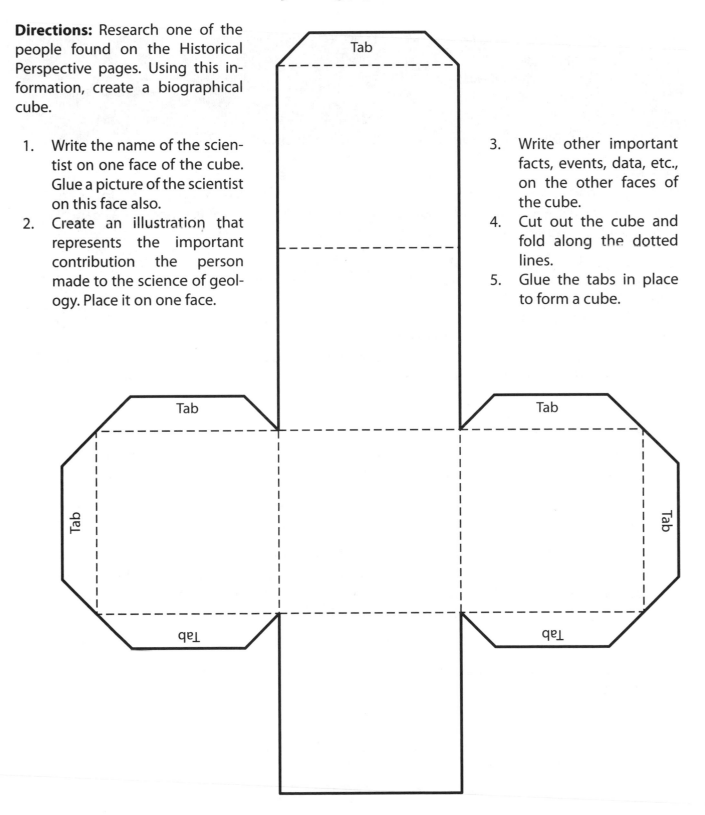

Unit 2: Layers of the Earth
Teacher Information

Topic: The earth is made of distinct layers.

Standards:
    NSES Unifying Concepts and Processes, (D)
    NCTM Geometry and Measurement
    STL Abilities for a Technological World
    See **National Standards** section (pages 61–65) for more information on each standard.

Concepts:
- The solid earth is layered with a lithosphere; a hot, convecting mantle; and a dense, metallic core.

Naïve Concepts:
- Earth is molten, except for its crust.
- The earth is round like a pancake.

Science Process Skills:
    Students will **identify** the layers of the earth. They will **describe** physical characteristics of the individual layers of the earth. Students will **compare and contrast** the layers of the earth. Students will **use a model to demonstrate** the layers of the earth.

Lesson Planner:
1. <u>Directed Reading</u>: Introduce the concepts and essential vocabulary relating to the layers of the earth using the directed reading exercise found on the Student Information pages.

2. <u>Assessment</u>: Evaluate student comprehension of the information in the directed reading exercise using the quiz located on the Quick Check page.

3. <u>Concept Reinforcement</u>: Strengthen student understanding of concepts with the activities found on the Knowledge Builder page. **Materials Needed:** Activity #1—12-inch Styrofoam ball, markers, serrated knife or jigsaw, legal-sized manila folders, scissors, stapler, rubber band; Activity #2—white paper, scissors, pen or pencil

Extension: Students create a poster identifying the layers of the earth, their compositions, and relative thicknesses.

Real World Application: Scientists are able to study the layers of the earth by analyzing the seismograms from earthquakes.

Unit 2: Layers of the Earth
Student Information

The earth is not a perfect sphere. It is slightly flattened at the poles and bulges slightly around the equator. Earth has a rocky surface covered with a thin layer of soil, but beneath this solid surface, Earth is very different. The earth consists of three main layers: crust, mantle, and core. Each layer has distinct characteristics.

The outermost layer of the earth is a relatively thin **crust**, also called the **lithosphere**. Some of Earth's crust is made up of soil. Beneath this soil is a thick layer of rock. Most rocks are made of the elements silicon and oxygen. The thickness of the crust ranges from 5 km (3 miles) to 100 km (62 miles).

Directly below the lithosphere is the **mantle**. The mantle is a hot, plastic-like layer of melted rock that surrounds the core. The mantle is about 2,970 km (1,856 miles) thick. It is also known as the **asthenosphere**. According to the Theory of Plate Tectonics, the upper mantle provides the basis upon which Earth's plates slide.

The center of the earth is the **core**. The core consists of two distinct layers. The liquid, outer core is made up mostly of iron and nickel with a thickness of 2,270 km (1,411 miles). The solid, inner core is made of iron and nickel with a thickness of 1,216 km (756 miles).

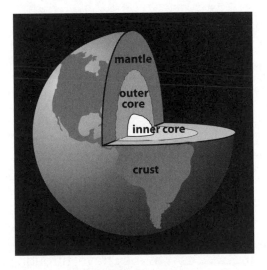

The earth is roughly spherical in shape. Layers of the earth are similar to the layers of a hardboiled egg. The brittle shell represents the lithosphere; the egg white represents the mantle; and the egg yolk represents the core.

Scientists make inferences about the interior of the earth based upon seismic waves. For example, certain seismic waves vary in speed as they travel through liquids and solids. Indirect observations from seismic data allow scientists to infer the composition (liquid or solid) and density of layers of the earth.

Name: _____ Date: _____

Quick Check

Matching

_____ 1. lithosphere
_____ 2. asthenosphere
_____ 3. crust
_____ 4. mantle
_____ 5. core

a. also called the lithosphere
b. a layer of dense, melted rock
c. outermost layer of the earth
d. plastic-like, upper mantle
e. innermost layer of earth

Fill in the Blanks

6. The mantle is a hot, plastic-like layer of melted _____.

7. Directly below the lithosphere is the _____, a layer of dense, molten rock that is about 2,970 km (1,856 miles) thick.

8. The core consists of two distinct sub-layers: the liquid, _____ _____ with a thickness of 2,270 km (1,411 miles), and the solid, _____ _____ with a thickness of 1,216 km (756 miles).

9. According to the Theory of _____ _____, the upper mantle provides the basis upon which Earth's plates slide.

10. Scientists make inferences about the interior of the earth based upon _____ _____.

Multiple Choice

11. Most rocks are made of _____ and oxygen.

    a. water
    b. nitrogen
    c. carbon
    d. silicon

12. Certain seismic waves vary in _____ as they travel through liquids and solids.

    a. size
    b. color
    c. speed
    d. shape

13. This hot, plastic-like layer of melted rock surrounds the core.

    a. asthenosphere
    b. lithosphere
    c. crust
    d. core

Name: _____   Date: _____

Knowledge Builder

Activity #1: Model of the Layers of Earth

Directions: Using markers, color and illustrate a Styrofoam ball to represent the earth. Cut the Styrofoam ball in half, creating two half spheres. Balls can easily be halved using a serrated knife or jigsaw, if available. (Teachers may wish to have the balls pre-cut for safety.) Illustrate and label the layers of the earth on each half of the ball. When the construction is complete, use a rubber band to hold the model halves together. Construct a base for the model by cutting strips (18 in. x 12 in.) from legal-sized manila folders. Staple the two ends together to form a collar on which to place the model.

Activity #2: Flapper Study Aide

Directions: Fold a piece of white paper in half like a hotdog bun. Turn the paper and fold it in fourths. Unfold the paper and cut up the fold making four flaps. On the fronts of the four flaps, write the layers of the earth: crust, mantle, outer core, and inner core. Underneath each flap, write important facts about each layer.

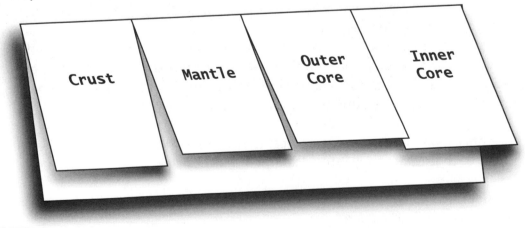

Unit 3: Rocks and Minerals
Teacher Information

Topic: There is a difference between rocks and minerals.

Standards:
>   **NSES** Unifying Concepts and Processes, (D)
>   **NCTM** Geometry, Measurement, and Data Analysis and Probability
>   **STL** Abilities for a Technological World
>   See **National Standards** section (pages 61–65) for more information on each standard.

Concepts:

- Rocks are mixtures that are usually made up of an assortment of minerals.
- A mineral is a naturally occurring, inorganic (nonliving) solid with a crystalline structure.

Naïve Concepts:

- Rocks, minerals, stones, and pebbles are all the same thing.

Science Process Skills:

> Students will be **manipulating** materials to make **predictions** and **conducting an experiment** to determine the physical characteristics/properties of rocks and minerals. Students will be **communicating** and **developing vocabulary** during the process of **collecting, recording, analyzing,** and **interpreting** data. Based on their findings, students will make an **inference** that there is a difference between rocks and minerals.

Lesson Planner:

1. <u>Directed Reading</u>: Introduce the concepts and essential vocabulary relating to rocks and minerals using the directed reading exercise found on the Student Information page.
2. <u>Assessment</u>: Evaluate student comprehension of the information in the directed reading exercise using the quiz located on the Quick Check page.
3. <u>Concept Reinforcement</u>: Strengthen student understanding of concepts with the activities found on the Knowledge Builder page. **Materials Needed:** Activity #1—sugar, salt, black paper, hand lenses, 3 x 5 index cards; Activity #2—an assorted selection of rocks and minerals, sandwich bags, markers, hand lenses
4. <u>Inquiry Investigation</u>: Explore rocks and minerals. Divide the class into teams. Instruct each team to complete the Inquiry Investigation pages. **Teacher Directions:** Provide each team a plastic baggie containing the following minerals: calcite, galena, quartz, magnetite, and talc. Set up mineral test stations around the room.

Extension: Students list all the man-made objects they can see in the classroom. They research the mineral makeup of each product they listed. Example: chalk: calcite.

Real World Application: Many fruits and vegetables are good sources of potassium, magnesium, and other minerals that our bodies need.

Unit 3: Rocks and Minerals
Student Information

Rocks are the basic material that makes up the earth's crust. **Rocks** are made up of minerals. Most rocks have several types of minerals in them. Granite is made of several minerals, including quartz, feldspar, and hornblende. Occasionally, a rock consists of only one mineral. Limestone is made of one mineral, calcite. Rocks may contain organic materials such as coal. Rocks may have various chemical compositions due to the presence of different minerals. Rocks are classified into three main groups: igneous, metamorphic, and sedimentary. These classifications are directly related to the processes under which the rocks are formed.

All rocks are made of minerals. A **mineral** is a naturally occurring, inorganic (nonliving) solid. It has a crystalline structure. This means that the atoms or ions that make up a mineral are arranged in an orderly and repetitive manner. It's made up of only ONE thing. A diamond is a mineral. It has a crystalline structure; it is made of one thing, carbon. Crystals form in one of six distinct shapes. A mineral has a crystal structure even if it does not have a crystal shape that you can see.

Minerals are the building blocks for rocks. Minerals combine differently to make rocks. Minerals are usually elemental compounds in their pure form. Minerals have four common features: minerals occur naturally, minerals are inorganic, minerals are solids, and minerals have a single chemical composition/structure. Minerals are relied upon as a major resource today. Most likely, if something isn't obtained from a plant or animal, it came from a rock or mineral.

Although there are over 4,000 known minerals, with new ones being discovered each year, only about 10 to 12 are abundant. Together, these abundant minerals account for what are known as the **rock-forming minerals**. The rock-forming minerals consist of 8 to 10 major elements, including: oxygen (O), silicon (Si), iron (Fe), aluminum (Al), calcium (Ca), sodium (Na), potassium (K), and magnesium (Mg). Most of the earth's crust is made of these major elements in some form.

Minerals can easily be classified into groups known as **families**. For example, one family of rock-forming minerals is known as the silicates. Silicates have a common crystalline structure, known as the **silicon-oxygen tetrahedron**. Quartz, a silicate, is a mineral found in many rocks on the earth's surface. In its pure form, quartz consists entirely of silicon-oxygen tetrahedral. Most other silicates combine with other elements, such as iron, sodium, calcium, potassium, and magnesium. Other families of minerals include: carbonates, halides, oxides, sulfides, and sulfates. These are considered to be non-silicates and account for only about one-quarter of the earth's continental crust.

Mineralogists commonly use a battery of tests to determine the identity of minerals. Although there are many tests for mineral identification, only seven of the most common will be examined. These are streak/color, hardness, luster, cleavage, specific gravity, magnetism, and reaction to acid.

Streak Color—Streak color should not be confused with the color of the mineral. It is possible for the color of a mineral to vary from color of the streak it leaves when dragged across a piece of white porcelain plate. The absence of streak should be noted as well; this information should be considered closely with the results of the hardness test.

Hardness—A test for how hard a mineral is; the Mohs' scale consists of ten minerals of varying hardness. The hardness of a mineral can be determined by rubbing an unknown mineral against the known minerals.

Luster —A test for how light interacts with the surface of the mineral. Observe minerals for one of the following classifications:
- Dull–no reflection of any kind
- Earthy–looks like dirt or dried mud
- Fibrous–appears to have fibers
- Greasy–looks like grease; may even feel greasy
- Metallic–perhaps shiny with the look of metals
- Pearly–looks like a pearl
- Silky–looks silky; sometimes hard to differentiate between fibrous; more compact
- Vitreous–the most common luster; the look of glass
- Waxy–looks like wax

Cleavage—This test is related to the crystalline structure and chemical bonding of the mineral and is determined by close observation of breaks within the sample. Minerals that are considered to have good cleavage tend to split/break uniformly along planes. For example, micas are made of tetrahedral sheets and break into thin sheets along the lines of the bonds. Galena tends to break into cubes with straight sides. These minerals have good cleavage. Many minerals do not have good cleavage; these are considered to be uneven or irregular and appear to be fractured along uneven planes/lines. In order to preserve the samples, you should not try to break minerals in any way.

Specific Gravity—This test refers to the weight of an amount of mineral compared to an equal amount of water. The specific gravity of the mineral is sometimes estimated via heft, e.g., lift equal amounts of quartz and galena. Simple water displacement can be used to determine the density of each mineral sample. Most minerals have a density of 2–3 g/cubic cm, but the heavier minerals often have densities as high as 7–20 g/cubic cm. Therefore, density will be more helpful in identifying heavier minerals.

Magnetism—Some minerals contain magnetic properties; this is easily tested by placing the mineral near a magnet.

Reaction to Acid—Minerals that react (usually fizzing or bubbling) tend to come from the carbonate family. While wearing safety glasses, place a small drop of vinegar on the mineral and observe carefully for any reaction.

Name: _____ Date: _____

Quick Check

Matching

_____ 1. luster   a. made up of minerals

_____ 2. Mohs' scale   b. a naturally occurring, inorganic (nonliving), solid

_____ 3. rocks   c. form in one of six distinct shapes

_____ 4. minerals   d. a test for how light interacts with the surface of the
                                           mineral

_____ 5. crystals   e. consists of ten minerals of varying hardness

Fill in the Blanks

6. Although there are over 4,000 known _____, with new ones being discovered
    each year, only about 10 to 12 are abundant.

7. Silicates have a common crystalline structure, known as the silicon-oxygen _____.

8. Minerals can easily be classified into groups known as _____.

9. The rock-forming minerals consist of 8 to 10 major _____, including: oxygen
    (O), silicon (Si), iron (Fe), _____ (Al), calcium (Ca), sodium (Na), potassium (K),
    and magnesium (Mg).

10. Rocks are classified into three main groups: igneous, _____, and
    _____.

Multiple Choice

11. The basic material that makes up the earth's crust is _____.
    a. mineral   b. limestone
    c. calcite   d. rock

12. Calcite is the only mineral that makes up _____.
    a. limestone   b. silicon
    c. granite   d. coal

13. Mineralogists commonly use a battery of tests to determine the identity of _____.
    a. rocks   b. coal
    c. minerals   d. diamonds

Name: _____   Date: _____

Knowledge Builder

Activity #1: Salt and Sugar Crystals

Directions: Examine the crystal shape of sugar and salt. Sprinkle sugar on black paper. Examine the crystals with a hand lens. Compare the shape of the sugar crystals to the six basic crystal shapes below. Now repeat the procedure with salt.

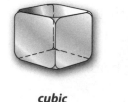

| **cubic** | **hexagonal** | **monoclinic** | **orthorhombic** | **tetragonal** | **triclinic** |

Observation

1. What is the shape of the sugar crystals? _____

2. What is the shape of the salt crystals? _____

Activity #2: Rocks vs. Minerals

Directions: Take two bags and label one "rocks" and the other "minerals". On 3 x 5 index cards, create two specimen description cards. On one card, write "Appears to be made of more than one substance." Place this in the rocks bag. On the other card, write "Appears to be made of only one substance." Place this in the minerals bag. Using a magnifying glass, examine and classify the

assortment of rock and mineral specimens given to you by your teacher. Place the samples in the correct bags. When classifying your specimens, refer to the data table below for help.

Rocks	**Minerals**
Rocks are usually made up of an assortment of materials. Rocks may contain organic materials. Rocks may have various chemical compositions due to the presence of more than one mineral.	Minerals occur naturally. Minerals are inorganic. Minerals are solids. Minerals have a single chemical composition/structure.

Name: _____ Date: _____

Inquiry Investigation: Mineral Identification

Concept:
- Minerals can be identified according to their physical properties: streak color, hardness, luster, cleavage, specific gravity, magnetism, and reaction to acid.

Purpose: Identify minerals by conducting various tests of their physical properties.

Procedure: Carry out the investigation. This includes gathering the materials, following the step-by-step directions, and recording the data.

Materials:

small white porcelain plates, unglazed   hand lenses
glass plates (caution: be sure the edges are ground)   water and graduated cylinders
pennies   steel file
magnets   safety goggles
samples of various common minerals   egg carton

Experiment:

Step 1: Take the minerals from the plastic baggie and place each one in a cup of the egg carton.

Step 2: Minerals are found in a variety of colors due to the chemicals in them. Record the color of each mineral in the data table.

Step 3: The amount of light reflected from the mineral's surface Is its luster. Luster is described as glassy, metallic, shiny, dull, waxy, satiny, or greasy. Record your description of the luster of each mineral in the data table.

Step 4: Some minerals leave a colored streak when rubbed across a piece of unglazed white tile. Rub a corner of each mineral across a piece of unglazed white tile several times to see if it will leave a streak. Some streak colors are different from the color of the mineral. If the mineral is 6.5 or harder, it will not leave a streak color. Use a hand lens to examine the streak plate to find the exact color. Record the color left by each mineral in the data table.

Step 5: Texture is the "feel" of the mineral's surface when it is rubbed. The feel is described as rough, smooth, bumpy, or soapy.

Step 6: All minerals are hard; the surface varies in resistance to scratching. Mohs' hardness scale of 1–10 is used to determine a mineral's hardness. The test is done with common materials such as fingernail, penny, steel nail, and glass. The hardness number is assigned based upon which item will scratch the mineral's surface. Test each mineral. It is not necessary to scratch the mineral with all the tools. The first tool to scratch the surface will determine the hardness. A hand lens will be helpful in seeing the surface more clearly to check for scratches. Use the Hardness Test Chart and Mohs' Hardness Scale to determine the hardness. Record the hardness number for each mineral in the data table.

Step 7: Check for magnetic properties by placing each mineral near a magnet. Record the results in the data table.

Name: _____ Date: _____

Hardness Test	
Minerals Scratched By	**Hardness Number**
fingernail	2.5 or less
penny	3 or less
glass (streak plate)	5.5 or less
steel nail	6.5 or less
none of the above	Greater than 6.5

Mohs' Hardness Scale

Mineral Examples	
Mineral	**Hardness**
Talc	1 (softest)
Gypsum	2
Calcite	3
Fluorite	4
Apatite	5
Feldspar	6
Quartz	7
Topaz	8
Corundum	9
Diamond	10 (hardest)

Results: Complete each test and record the information in the data table below. Compare your results with the Mineral Identification Key below. Find the name of your mineral by matching your mineral's information to the descriptions in the key. Record the name of each mineral in the data table below.

Mineral Test Data Table

Rock	Color	Luster	Texture	Streak Color	Mohs' Hardness Number	Magnetic	Mineral Name
#1							
#2							
#3							
#4							
#5							

Mineral Identification Key

Directions: To identify your minerals, match your descriptions to the ones in the data table below.

Color	Luster	Texture	Streak Color	Hardness Number	Magnetic	Mineral Name
silver & shiny	metallic	smooth	white or pink	3	no	galena
pink, purple, white	shiny & glassy	smooth to bumpy	white	6 to 7	no	quartz
grey or black	dull	rough	black or dark grey	7	yes	magnetite
light grey	dull	feels like soap	white	1	no	talc
tan & white	shiny & glassy	smooth	white or pink	3	no	calcite

Conclusion:

Explain how minerals can be identified by conducting various tests of their physical properties.

_____

_____

_____

Unit 4: Plate Tectonics
Teacher Information

Topic: Plate tectonics explains how the continents were able to move to their present locations.

Standards:
NSES Unifying Concepts and Processes, (D)
NCTM Geometry and Measurement
See **National Standards** section (pages 61–65) for more information on each standard.

Concepts:
- The crust of the Earth has fractured into seven major tectonic plates that sometimes collide and grind past each other, causing earthquakes, volcanic activity, mountain-building, and oceanic trench formation.

Naïve Concepts:
- Earthquakes and volcanic activity are random occurrences.
- All the mountains were formed when the earth was created.

Science Process Skills:
  Students will be **communicating** and **developing vocabulary** during the process of **collecting, recording, analyzing,** and **interpreting** data. Based on their findings, students will make an **inference** that earthquakes, volcanic activity, mountain-building, and oceanic trench formation occur along plate boundaries.

Lesson Planner:
1. Directed Reading: Introduce the concepts and essential vocabulary relating to plate tectonics using the directed reading exercise found on the Student Information pages.
2. Assessment: Evaluate student comprehension of the information in the directed reading exercise using the quiz located on the Quick Check page.
3. Concept Reinforcement: Strengthen student understanding of concepts with the activities found on the Knowledge Builder page. **Materials Needed:** Activity #1—white paper, scissors; Activity #2—hard-boiled egg, markers, plastic knife

Extension: Students keep track of earthquakes and volcanoes that occur throughout the world for a year on a world map. Students determine which parts of the world seem most geologically active.

Real World Application: An ocean-bottom quake west of Indonesia on December 26, 2004, created a tsunami that killed at least 145,000 coastal residents and tourists. As a result of this tragedy, early warning systems for tsunamis have been developed.

Unit 4: Plate Tectonics
Student Information

Plate Tectonics and Continental Drift are both terms used to explain the present location of the continents. **Continental Drift** is the theory, proposed in the early 1900s, that states a supercontinent called **Pangaea** broke apart, forming the seven continents. The continents slowly drifted to their present positions. Today scientists use **Plate Tectonics** to explain how the continents were able to move to their present locations. Scientists believe the earth's outermost layer, the lithosphere, broke apart and formed seven large plates (pieces) and many smaller plates. The motion of magma in the mantle or asthenosphere, just under the crust, causes the movement of the plates. Scientists believe the seven continents were once part of a supercontinent called Pangaea.

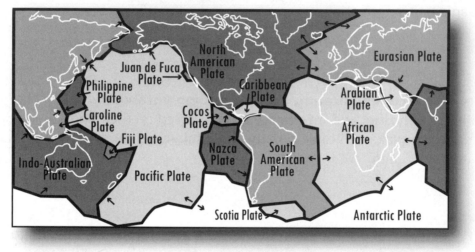

The plates move between 5 and 10 cm a year. The plates sometimes collide and grind past each other at the plate boundaries. Earthquakes, volcanic activity, mountain-building, and oceanic trench formation occur along plate boundaries. Three types of plate boundaries are convergent boundaries, divergent boundaries, and transform boundaries. **Convergent plate boundaries** crash together, often forming mountains like the Himalayas and producing earthquake and volcanic activity. **Divergent plate boundaries** move apart. Most divergent boundaries are under the ocean. They build undersea mountain ranges called mid-ocean ridges. **Transform boundaries** form where two plates slide past each other, often causing earthquakes.

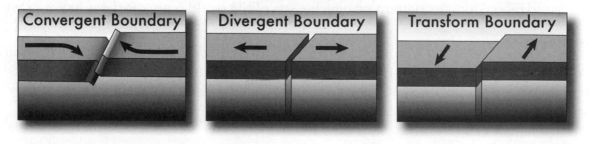

Pressure builds up when two plates meet. This is known as a **fault line**. When too much pressure builds up, the rocks suddenly slide past each other and the pressure is released. The result is an earthquake. The **focus** is the place in the earth's crust where the pressure was released. The focus can be many kilometers below in the crust. Earthquake waves spread out in all directions from the focus. The earthquake's **epicenter** is the spot on the earth's surface directly above the focus. Quakes at sea can displace tons of water and create waves called **tsunamis** that create havoc when they reach land. The ocean-bottom quake west of Indonesia on December 26, 2004, created the tsunami that killed at least 145,000 coastal residents and tourists.

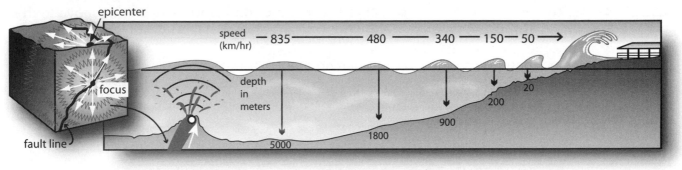

Movement in the earth's crust as a result of the plate boundaries colliding and grinding past each other creates mountains. Mountains are classified by the way they are formed. **Folded mountains** are formed when the earth's crust folds into great waves. The Rocky Mountain Range was formed by the folding of the earth's curst. **Faulted mountains** are formed when the earth's crust breaks into huge blocks. Some blocks move upward and some downward. The Teton Mountain Range is an example of faulted mountains. **Volcanic mountains** are formed when converging boundaries crash together. The heat and pressure becomes so great that rock melts. The molten rock is forced to the surface of the earth, forming mountains. Volcanoes can be classified into three groups: cinder cone volcanoes, shield volcanoes, and composite cone volcanoes. Cinder cones are built up when cinders and ash are ejected from the volcano. Shield volcanoes form as layers of lava ooze out of the volcano and cool. Composite volcanoes are a combination of cinders and lava flows.

Folded Mountains

Faulted Mountains

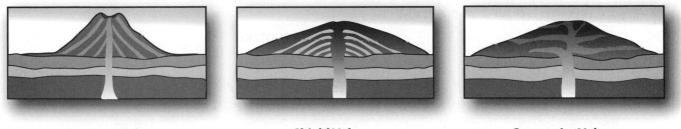

Cinder Cone Volcano   **Shield Volcano**   **Composite Volcano**

Name: _____   Date: _____

Quick Check

Matching

_____ 1. Pangaea

_____ 2. Plate Tectonics

_____ 3. focus

_____ 4. epicenter

_____ 5. folded mountains

a. place in the earth's crust where the pressure was released

b. the spot on the earth's surface directly above the focus

c. formed when the earth's crust folds into great waves

d. explains how the continents were able to move to their present locations

e. supercontinent

Fill in the Blanks

6. _____ _____ are formed when the earth's crust breaks into huge blocks.

7. _____ _____ are formed when converging boundaries crash together.

8. _____ _____ is the theory proposed in the early 1900s that a supercontinent called Pangaea broke apart, forming the seven continents.

9. _____ _____ and _____ _____ are both terms used to explain the present location of the continents.

10. Scientists believe the earth's outermost layer, the _____, broke apart and formed seven large plates (pieces) and many smaller ones.

Multiple Choice

11. These boundaries crash together, often forming mountains like the Himalayas and producing earthquake and volcanic activity.

    a. transform boundaries

    c. convergent plate boundaries

    b. divergent plate boundaries

    d. supercontinent boundaries

12. Most divergent boundaries are in the _____.

    a. oceans

    c. plains

    b. mountains

    d. valleys

13. Quakes at sea can displace tons of water and create waves called _____.

    a. tidals

    c. focus

    b. tsunamis

    d. epicenters

Name: _____ Date: _____

Knowledge Builder

Activity #1: Three-Tab Cause and Effect Book

Directions: Fold a piece of 8 1/2" X 11" paper vertically, like a hotdog bun. Fold it in thirds horizontally, like a hamburger. Unfold the paper, leaving it still folded in the hotdog bun position. Starting at the edge of the paper, cut up the crease line folds to the top of the hotdog fold. You should have 3 lift tabs. On the top of the first tab, write the word "Cause" and just beneath the word, cause, write the word, "Convergent Boundary." On the middle tab, write the word "Cause" and beneath it, write "Divergent Boundary." On the third tab, write the word "Cause" and below it, write "Transform Boundary." Now, lift the tabs and write the word, "Effect" at the top of each. Write and illustrate what happens (effect) along each of these boundaries.

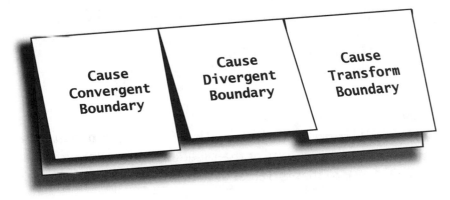

Activity #2: Egg Tectonics

Directions: In this activity, the hard-boiled egg will represent Earth. Gently tap the egg on a hard surface to create as few cracks as possible. Try to make the cracks large fragments; these fragments will represent the earth's plates. With a colored marker, trace the major cracks. The edges of the cracks represent the earth's different tectonic plates. Now, squeeze the egg gently to create a slight movement of the shell pieces. Places where the eggshells separate represent a divergent boundary. Where two pieces of eggshell come together, a convergent boundary is represented. A transform boundary is represented when one piece of eggshell slides past the other. What geological activities might occur along these boundaries?

1. Divergent boundary _____

   _____

2. Convergent boundary _____

   _____

3. Transform boundary _____

   _____

Unit 5: Soil, Weathering, and Erosion
Teacher Information

Topic: The crust of the earth is constantly changing.

Standards:
>   **NSES** Unifying Concepts and Processes, (D)
>   **NCTM** Measurement and Data Analysis and Probability
>   **STL** Abilities for a Technological World
>   See **National Standards** section (pages 61–65) for more information on each standard.

Concepts:
- Soil comes from weathered rock fragments.
- Weathering changes the earth's surface over time by breaking rock into smaller pieces.
- Erosion is the wearing away of the earth's surface by wind, water, ice, or gravity.

Naïve Concepts:
- Soil is alive.
- Only the brown, loamy type soil found in flower beds is true soil.
- Weathering and erosion are essentially the same thing.

Science Process Skills:

Students will be **collecting, recording,** and **interpreting information** while **developing the vocabulary to communicate** the results of their reading and exploration. Based on their findings, students will make an **inference** that the crust of the earth is constantly changing.

Lesson Planner:
1. Directed Reading: Introduce the concepts and essential vocabulary relating to soil, weathering, and erosion using the directed reading exercise found on the Student Information page.
2. Assessment: Evaluate student comprehension of the information in the directed reading exercise using the quiz located on the Quick Check page.
3. Concept Reinforcement: Strengthen student understanding of concepts with the activities found on the Knowledge Builder page. **Materials Needed:** Activity #1—four types of soil (lawn soil, potting soil, sandy soil, poor soil), hand lens, Styrofoam cups, beaker, graduated cylinder, water, pencil; Activity #2—straight-sided baking pan or wooden frame resting on a pane of glass, plaster of Paris, water, wet paper towels, petroleum jelly, sprouted peas or beans, terrarium

Extension: Students explore the school grounds for evidence of the transportation of soil by water. Sediment is often deposited along walks and driveways and in low places on the lawn.

Real World Application: In the 1930s, rolling dust storms swept across the Great Plains with devastating effects on the people and land. Drought, combined with prior mistreatment of the land, led to one of the greatest natural disasters in the United States, the Dust Bowl.

Unit 5: Soil, Weathering, and Erosion
Student Information

Soil is a valuable natural resource. All life on land depends on the soil. Plants are rooted in the soil and obtain nutrients from it for growth. Animals eat plants or other animals that eat plants.

Soil is a mixture of nonliving things, such as sand grains, smaller rock particles, and minerals. It contains organic material that comes from decaying dead plants and animals. It also holds living things, both plants and animals. There are many different types of soil, and each one has unique characteristics. The kinds of soils in an area help determine how well crops grow there. Soil can be acid, alkaline, or neutral. Highly acidic or alkaline soils can harm many plants.

Soil Characteristics

Characteristic	Description
Color: depends on the amount of air, water, organic matter, and certain elements in the soil	<u>Brown to black</u>: accumulation of organic matter, humus <u>Purplish-black</u>: accumulation of manganese <u>Yellow to reddish</u>: accumulation of iron <u>White to gray</u>: accumulation of salt
Texture: determines how well water drains from a soil. Sands promote drainage better than clays.	<u>Sandy</u>: feels rough <u>Silt</u>: feels soft, silky, or floury <u>Clay</u>: smooth when dry and sticky when wet
Structure: the arrangement of smaller soil particles (sand, silt, and clay) to form larger pieces	<u>Granular</u>: individual particles of sand, silt, and clay grouped together in small, round grains <u>Blocky</u>: soil particles cling together in block shapes <u>Prismatic</u>: soil particles have formed into vertical pillars <u>Platy</u>: soil particles form thin sheets piled horizontally on one another
Chemical Condition: influences what grows and lives in the soil	Soil is measured by pH values of 1–14: <u>neutral</u>: pH of 7 <u>acidic</u>: pH below 7 <u>alkaline</u>: pH above 7

Soil comes from weathered rock fragments. It has taken thousands of years to form. Soil is a combination of rock, mineral particles, and inorganic matter. It forms distinct layers known as **horizons**. Mature soil has four layers, and each layer has its own special content of minerals, color, and texture. The uppermost layer contains decaying leaves and animal remains needed to form the loose, rich topsoil just beneath it. Below the topsoil horizon lies the subsoil. It contains minerals washed down from the topsoil horizon. Humus, clay, and tiny soil particles are also in this layer. The fourth horizon, located beneath the subsoil, holds mostly weathered rock pieces.

The crust of the earth is constantly changing. Movement in the earth's crust as a result of the plate boundaries colliding and grinding past each other creates mountains. Weathering changes the earth's surface over time by breaking rock into smaller pieces. Through the process of erosion, soil and other rock by-products are swept away and deposited in other places.

There are two types of weathering: chemical and mechanical. **Chemical weathering** causes changes in the chemical makeup of rocks and makes them crumble. Chemical weathering in caves causes stalactites and stalagmites. **Mechanical weathering** is the breaking of rock into smaller pieces by physical forces such as wind, water, plants, and freezing. Mechanical weathering can be seen when expanding ice breaks a rock into smaller pieces.

The small pieces of rock broken down by weathering can be carried away by erosion. **Erosion** is the wearing away of the earth's surface by wind, water, ice, or gravity. Erosion takes away the soil in one place and deposits it in another. When these materials are swept to a new location, it is called **deposition**. Over time, moving water creates an eroded path, causing gullies and canyons. Rain carries sediment from the land to larger bodies of water. The sediments form rich **deltas** at the mouths of rivers before entering the ocean. Planting trees and grasses can help soak up and slow down the water, and this helps slow erosion.

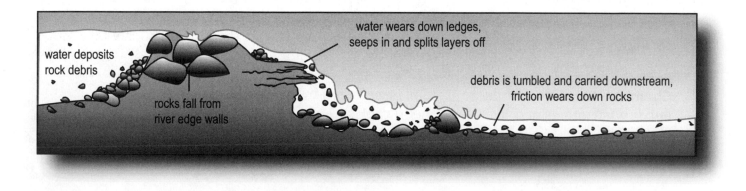

water deposits rock debris

rocks fall from river edge walls

water wears down ledges, seeps in and splits layers off

debris is tumbled and carried downstream, friction wears down rocks

Name: _____ Date: _____

Quick Check

Matching

_____ 1.   erosion

_____ 2.   chemical weathering

_____ 3.   deltas

_____ 4.   mechanical weathering

_____ 5.   soil

a.   breaking of rock into smaller pieces by physical forces such as wind, water, plants, and freezing

b.   wearing away of the earth's surface by wind, water, ice, or gravity

c.   causes changes in the chemical makeup of rocks

d.   a combination of rock, mineral particles, and inorganic matter

e.   rich sediment formations at the mouths of rivers

Fill in the Blanks

6.   _____ comes from weathered rock fragments.

7.   Soil is a mixture of nonliving things, such as sand grains, smaller rock particles, and _____.

8.   Soil can be acid, _____, or neutral.

9.   Soil is measured by _____ values of 1–14.

10. The _____ of the earth is constantly changing.

Multiple Choice

11. Soil forms distinct layers called _____.

   a.   plate boundaries   b.   horizons

   c.   stalagmites   d.   characteristics

12. This determines how well water drains from a soil.

   a.   texture   b.   structure

   c.   color   d.   chemical condition

13. Soil with a pH below 7 is called _____.

   a.   alkaline   b.   neutral

   c.   salty   d.   acidic

Name: _____   Date: _____

Knowledge Builder

Activity #1: Soil Test

Directions: Examine four different types of soil (lawn soil, potting soil, sandy soil, poor soil) samples. Complete each soil test and record the information in the data table below.

Test #1—Soil Color: Examine the soil samples and record their color in the table below.
Test #2—Soil Composition: Examine soil samples using a magnifying glass. Identify any decaying plant or animal matter, debris, and/or living organisms found in the samples. Record the data in the table below.
Test #3—Soil Texture: Rub each soil sample between your fingers. Identify the texture, and record the data in the table below.
Test #4—Soil Water-Holding Capacity: Using a pencil, punch a small hole in the bottom of a Styrofoam cup. Add soil until the cup is half full. Hold the cup over a beaker. Add 100 ml of water to the graduated cylinder. Then pour the water over the top of the soil. In the table below, record the amount of time it takes the water to collect in the beaker.

Soil Test	Sample #1	Sample #2	Sample #3	Sample #4
Test #1—Color				
Test #2—Compostion				
Test #3—Texture				
Test #4—Water-Holding Capacity				

Activity #2: Effects of Plants on Rocks

Directions: Make a one-inch-thick slab of plaster of Paris either in a straight-sided baking pan coated with petroleum jelly or by pouring the plaster into a wooden frame resting on a pane of glass. After the plaster has hardened, remove it from the mold. Lay sprouted pea or bean seeds on the smooth side of the plaster and cover them with wet paper towels. Keep the slab (and seeds) in a terrarium so that the towels stay moist. After several days, remove the seeds.

Observation: What happened to the surface of the plaster? Why? _____

_____

_____

Conclusion: What effect do plants have on rocks? _____

_____

_____

Unit 6: Sedimentary Rocks
Teacher Information

Topic: Sedimentary rocks are the result of the process of physical and chemical forces in nature.

Standards:
> **NSES** Unifying Concepts and Processes, (D)
> **NCTM** Geometry and Measurement
> **STL** Abilities for a Technological World
> See **National Standards** section (pages 61–65) for more information on each standard.

Concepts:
- Sedimentary rock is formed by the weathering and erosion of existing rocks.

Naïve Concepts:
- Rocks grow.
- Sedimentary rocks form as puddles dry up.

Science Process Skills:

Students will **develop vocabulary** relating to sedimentary rocks. They will **identify** the main ingredients of sedimentary rocks. Students will **infer** the relationship between weathering and erosion and sedimentary rock formations.

Lesson Planner:
1. <u>Directed Reading</u>: Introduce the concepts and essential vocabulary relating to sedimentary rocks using the directed reading exercise on the Student Information pages.
2. <u>Assessment</u>: Evaluate student comprehension of the information in the directed reading exercise using the quiz located on the Quick Check page.
3. <u>Concept Reinforcement</u>: Strengthen student understanding of concepts with the activities found on the Knowledge Builder page. **Materials Needed:** Activity #1—sand, gravel, topsoil, clear peanut butter jar with lid, water; Activity #2—clean milk carton, sand, coarse sand, different-sized gravel, sawdust or fine bark chips, plaster of Paris mixed according to the directions on the package

Extension: Depending upon your geographical location, locate a road-cut that clearly shows sedimentary rock layers. Arrange for a field trip to the location. Perhaps arrange for a geologist to offer an interpretation of the rock formation.

Real World Application: Sedimentary rocks are the only rocks that contain fossils. Plants and animals that have died get covered up by new layers of sediment. They may be turned into stone, or they may leave impression in the rock. Most fossils are plants and animals that lived in the sea.

Unit 6: Sedimentary Rocks
Student Information

The minerals formed deep within the earth combine in various ways to form the hard solids we call rocks. **Sedimentary rock** is a kind of rock formed when a layer of sediment becomes solid. This type of rock is formed by the weathering and erosion of existing rocks. Rocks exposed to air and water slowly wear away. During the process of weathering, small pieces, or particles, break away from the main rock. Water flowing over the earth's surface picks up **sediment**, or small pieces of rock, sand, clay, and other materials. The water flows into streams or rivers. As the flow slows down, some particles of rock and other materials fall to the bottom of the river and settle out of the water.

Rock particles in flowing water settle out and layers form. The layers of sediment become covered by other layers. The upper layers press down on the lower layers. The weight of accumulated particles, along with mineral-laden water, cements everything together. After thousands of years, layers of sediment become solid and form rocks. Sandstone is an example of a sedimentary rock that was formed when layers of sediment became solid.

Sedimentary rocks are known for their layered characteristics and often provide clues to the past in the form of fossils. Sedimentary rocks are classified by the source of their sediments. There are three classifications of sedimentary rocks.

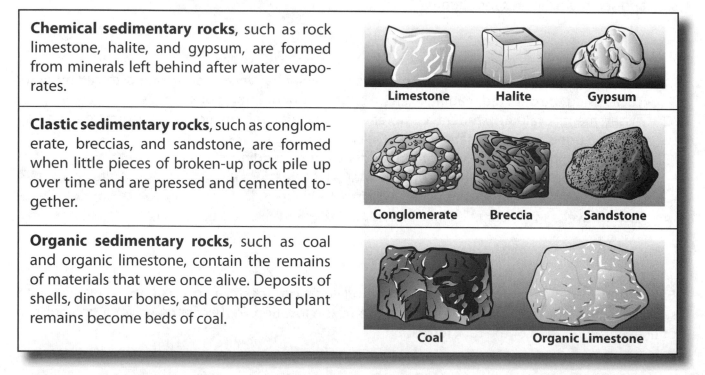

Chemical sedimentary rocks, such as rock limestone, halite, and gypsum, are formed from minerals left behind after water evaporates.	Limestone Halite Gypsum
Clastic sedimentary rocks, such as conglomerate, breccias, and sandstone, are formed when little pieces of broken-up rock pile up over time and are pressed and cemented together.	Conglomerate Breccia Sandstone
Organic sedimentary rocks, such as coal and organic limestone, contain the remains of materials that were once alive. Deposits of shells, dinosaur bones, and compressed plant remains become beds of coal.	Coal Organic Limestone

All sedimentary rocks are formed by two common processes: compaction and cementation. **Compaction** is caused by the weight of the overlying rocks squeezing the sediments together. **Cementation** is a chemical process through which water carries and deposits dissolved minerals in the small spaces between the sediment particles. In a sense, the particles are cemented together.

Name: _____ Date: _____

Quick Check

Matching

_____ 1. sedimentary rock

_____ 2. organic sedimentary rocks

_____ 3. sediment

_____ 4. chemical sedimentary rocks

_____ 5. compaction

a. contain the remains of materials that were once alive

b. formed from minerals left behind after water evaporates

c. caused by the weight of the overlying rocks squeezing the sediments together

d. kind of rock formed when a layer of sediment becomes solid

e. small pieces of rock, sand, clay, and other materials

Fill in the Blanks

6. _____ is a chemical process through which water carries and deposits dissolved minerals in the small spaces between the sediment particles.

7. All sedimentary rocks are formed by two common processes: _____ and _____.

8. The _____ formed deep within the earth combine in various ways to form the hard solids we call rocks.

9. During the process of _____, small pieces, or particles, break away from the main rock.

10. Sedimentary rocks are known for their layered characteristics and often provide clues to the past in the form of _____.

Multiple Choice

11. These rocks are formed when little pieces of broken-up rock that pile up over time are pressed and cemented together.
    - a. clastic sedimentary
    - b. chemical sedimentary
    - c. organic sedimentary
    - d. compact sedimentary

12. An example of an organic sedimentary rock is _____.
    - a. halite
    - b. coal
    - c. gypsum
    - d. sandstone

Name: _____   Date: _____

Knowledge Builder

Activity #1: Settle Down

Directions: Place equal amounts of sand, gravel, and topsoil, in an empty jar. The rocks and soils should take up roughly 60% of the total volume of the jar. Fill the remaining 40% of the jar with water. Fasten the lid tightly and shake vigorously for several minutes. Observe the contents of the jar for several days, until the water is clear. Sketch and describe the layers that formed in the jar.

Top: _____

_____

_____

Middle: _____

_____

_____

Bottom: _____

_____

_____

Activity #2: Make a Sedimentary Rock Model

Directions: Cut the top off of a clean milk carton. Place a layer of fine sand, coarse sand, different-sized gravel, and sawdust or fine bark chips in the carton. Make the layers different sizes. Continue layering the materials until the carton is filled. Mix the plaster of Paris according to the directions on the package. Pour the liquid into the milk carton until all layers of material are covered. Let the plaster set until dry. Tear away the milk carton. Examine your model of a sedimentary rock.

Conclusion: How is your sedimentary rock model similar to sedimentary rock found in nature?

_____

_____

_____

Unit 7: Igneous Rocks
Teacher Information

Topic: Igneous rocks are the result of the process of physical forces in nature.

Standards:
> **NSES** Unifying Concepts and Processes, (D)
> **NCTM** Geometry, Measurement, and Data Analysis and Probability
> **STL** Abilities for a Technological World
> See **National Standards** section (pages 61–65) for more information on each standard.

Concepts:
- Igneous rocks are formed when melted rock cools and hardens, either under or above the surface of the earth.

Naïve Concepts:
- Rocks grow.
- All rocks are the same, and it's hard to tell how they originated.

Science Process Skills:
> Students will **develop vocabulary** relating to igneous rocks. They will **identify** the main components of igneous rocks. Students will **infer** the relationship between plate tectonics and igneous rock formations.

Lesson Planner:
1. Directed Reading: Introduce the concepts and essential vocabulary relating to igneous rocks using the directed reading exercise found on the Student Information page.
2. Assessment: Evaluate student comprehension of the information in the directed reading exercise using the quiz located on the Quick Check page.
3. Concept Reinforcement: Strengthen student understanding of concepts with the activities found on the Knowledge Builder page. **Materials Needed**: Activity #1—A warm batch of Rice Krispy™ Marshmallow Treats (Follow the directions on the Rice Krispy™ cereal box.), spray cooking oil, raisins, chocolate chips, nuts, M&M's™, waxed paper, rolling pin; Activity #2—flour, salt, warm water, red food coloring, water, vinegar, baking soda, cardboard, large bowl, cookie sheet, access to an oven, paper cup, brown tempera paint, paint brush, scissors, glue, construction paper
4. Inquiry Investigation: Explore viscosity and how it relates to magma. Divide the class into teams. Instruct each team to complete the Inquiry Investigation pages.

Extension: Pumice, a type of igneous rock, is often used as a decorative landscape stone. Students research and compile a list of igneous rocks and their common uses.

Real World Application: In May 1980, Mount St. Helens erupted over the course of several days. The eruption is considered the deadliest and most economically destructive volcanic event in the history of the United States.

Unit 7: Igneous Rocks
Student Information

Igneous rock, known as "fire rock," is rock formed by the cooling of melted material, such as magma inside the earth and lava above the ground. Igneous rocks are formed as a result of activity at plate boundaries: volcanoes and sea-floor spreading. Obsidian is an example of igneous rock. Obsidian forms when lava cools quickly above ground.

Obsidian

Volcanoes are the result of converging boundaries crashing together. The heat and pressure becomes so great that rock melts within the upper mantle, or asthenosphere, layer of the earth. The molten rock is forced to the surface, forming mountains of rock. If enough pressure builds up, the molten rock may be forced out of the earth in an explosive eruption. Molten rock is known as **magma** and is converted into **lava** as it is **extruded** (forced through an opening in the earth) onto the earth's surface to cool. This process forms extrusive igneous rocks. Basalt is an example of an **extrusive igneous rock**. Magma that cools inside the earth forms **intrusive igneous rocks**. Granite is an example of an intrusive igneous rock.

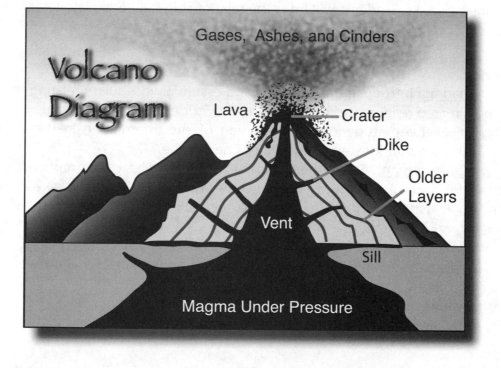

Physical properties of igneous rocks are largely determined by the rate at which the rocks cool and the manner in which they are extruded. Slowly cooling lava results in larger, more definitive crystals, whereas rapidly cooling crystals are often so small they can't be seen with the naked eye. Geologists use crystal size and arrangement as clues to identify the conditions under which igneous rocks were formed.

The mineral composition of igneous rocks depends on minerals in the magma. Igneous rocks are divided into three families depending on the minerals they contain.

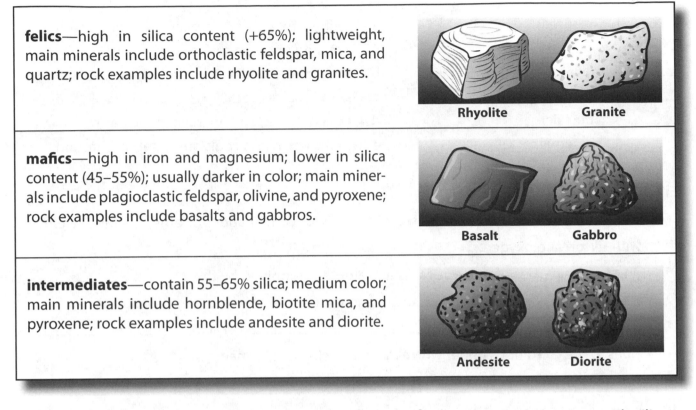

felics—high in silica content (+65%); lightweight, main minerals include orthoclastic feldspar, mica, and quartz; rock examples include rhyolite and granites.

Rhyolite   Granite

mafics—high in iron and magnesium; lower in silica content (45–55%); usually darker in color; main minerals include plagioclastic feldspar, olivine, and pyroxene; rock examples include basalts and gabbros.

Basalt   Gabbro

intermediates—contain 55–65% silica; medium color; main minerals include hornblende, biotite mica, and pyroxene; rock examples include andesite and diorite.

Andesite   Diorite

Magma is composed of **silicates**, or combinations of other chemical elements with silicon and oxygen. The amount of silica controls the **viscosity** (the resistance of a substance to flow). Magmas with high viscosity tend to contain higher amounts of silica and commonly form lighter-colored igneous rocks. The highly viscous magma often builds up under intense pressure and is associated with spectacular volcanic eruptions. Low viscosity basalts, which contain less silica, are commonly associated with sea-floor spreading as thin magma oozes through the earth's surface.

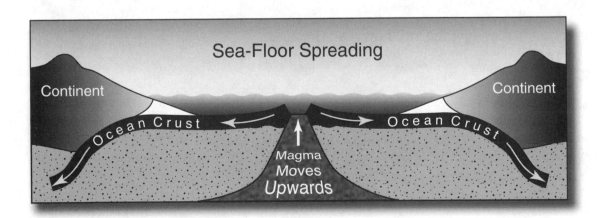

Sea-Floor Spreading

Continent   Continent

Ocean Crust   Ocean Crust

Magma Moves Upwards

Name: _____   Date: _____

Quick Check

Matching

_____ 1. magma

_____ 2. lava

_____ 3. extruded

_____ 4. obsidian

_____ 5. igneous rock

a. molten rock that has been extruded through an opening onto the surface of the earth

b. molten rock under the earth's surface

c. known as "fire rock"

d. forced through an opening in the earth

e. forms when lava cools quickly above ground

Fill in the Blanks

6. _____ is composed of silicates, or combinations of other chemical elements with silicon and oxygen.

7. Physical properties of igneous rocks are largely determined by the _____ at which the rocks _____ and the manner in which they are extruded.

8. Geologists use _____ _____ and _____ as clues to the conditions under which igneous rocks were formed.

9. The mineral composition of igneous rocks depends on _____ in the magma.

10. Igneous rocks are formed as a result of activity at _____ _____.

Multiple Choice

11. Magma that cools inside the earth forms _____.
    a. extrusive igneous rocks
    b. intrusive igneous rocks
    c. crystals
    d. minerals

12. Igneous rocks are divided into how many families?
    a. two
    b. five
    c. three
    d. four

13. Magmas with high viscosity tend to contain higher amounts of _____.
    a. silica
    b. iron
    c. magnesium
    d. pyroxene

Name: _____ Date: _____

Knowledge Builder

Activity #1: Edible Igneous Rock

Directions: Make a batch of Rice Krispy™ Marshmallow Treats (follow the directions on the Rice Krispy™ cereal box). Spray your hands with cooking oil. Place a cup of the warm Rice Krispy mixture on the waxed paper square. Sprinkle other ingredients, like raisins, chocolate chips, nuts, and M&M's™, on the mixture. Roll the mixture with a rolling pin sprayed with cooking oil. When cooled, examine your igneous rock and then eat it.

Conclusion: How is your edible rock similar to an igneous rock? _____

_____

_____

_____

Activity #2: Model of a Volcano

Directions: First, make a salt and flour volcano model. In a large bowl, place 1000 mL (1 L) of flour and 250 mL of salt. Gradually add 250 mL of warm water, and carefully mix together using your hands. Add more water as needed until the mixture becomes a doughy substance. Then place it on a cookie sheet and knead it until it is smooth and rubbery. Using the dough, mold it into a volcano. Leave an opening at the top of the cone deep enough to conceal a paper cup. Bake in the oven for 30 minutes at 300 degrees. Remove your volcano from the oven and allow it to cool. Place a paper cup inside the opening. Paint your volcano with brown tempera paint. Allow time for drying. Place your model volcano on top of a sheet of cardboard. Draw and cut out paper trees. Glue the trees to your cardboard at the base of the volcano to create a forest.

Second, prepare for a simulated eruption. In a paper cup, mix 45 mL of baking soda with 118 mL of water. Add red food coloring until mixture turns red. Pour the mixture into the cup at the top of your volcano. Place small bits of Styrofoam in the cup. Pour 5 mL of vinegar into the mixture. Observe the reaction.

Conclusion: How is your simulated eruption similar to a real volcanic eruption? _____

_____

_____

Name: _____   Date: _____

Inquiry Investigation: Viscosity

Concept:
- The resistance of a substance to flow is called **viscosity**. Substances with high viscosity flow slowly. Low-viscosity substances flow more quickly.

Purpose: Does temperature affect the viscosity (rate of flow) of a liquid?

Hypothesis: Write a sentence that predicts what your scientific investigation will prove.

_____

_____

Procedure: Carry out the investigation. This includes gathering the materials, following the step-by-step directions, and recording the data.

Materials:

9 x 13-inch cake pan	creamy peanut butter	molasses
Murphy's Oil Soap™	vegetable oil	honey
small condiment cups	5 x 7-inch sheets of cardboard	stopwatch
access to a microwave	microwave-safe bowls	hot pads

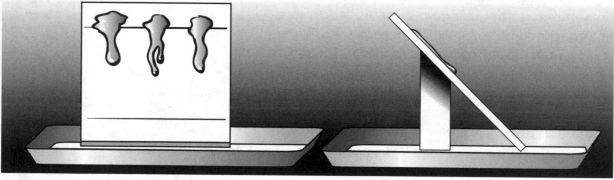

Front View   **Side View**

Experiment:

Controlled Setup

Step 1:   Fill a small condiment cup with room temperature peanut butter.

Step 2:   Place a cardboard sheet at a 45-degree angle within the cake pan (see diagram above).

Step 3:   Draw a starting point near the top of the card and an ending point near the bottom.

Step 4:   Drop the peanut butter onto the card. When the peanut butter touches the start Line, start timing.

Step 5:   Stop timing when the flow touches the stop line. Calculate the time it took the flow to reach the stop line. Record the time (seconds) in the data table.

Step 6:   Repeat Steps 1–5 two more times with the peanut butter.

Step 7:   Repeat Steps 1–6 for the molasses, Murphy's Oil Soap™, vegetable oil, and honey.

Name: _____   Date: _____

Results:

Record the time in seconds for each liquid tested in the data table below. Calculate and record the average time for each liquid in the data table.

Controlled Setup				
Room Temperature Liquid	**Trial #1 Flow Time (sec)**	**Trial #2 Flow Time (sec)**	**Trial #3 Flow Time (sec)**	**Average Flow Time (sec)**
peanut butter				
molasses				
Murphy's Oil Soap™				
vegetable oil				
honey				

Experimental Setup

Step 1:   Repeat Steps 1–6 for the peanut butter, molasses, Murphy's Oil Soap™, vegetable oil, and honey, except heat the liquids for 30 seconds in a microwave before pouring onto the cardboard. Be sure to use microwave-safe bowls to heat the liquids. Also, use hot pads to handle the bowls after they have been heated.

Results:

Record the time in seconds for each liquid tested in the data table below. Calculate and record the average time for each liquid in the data table.

Experimental Setup				
Heated Liquid	**Trial #1 Flow Time (sec)**	**Trial #2 Flow Time (sec)**	**Trial #3 Flow Time (sec)**	**Average Flow Time (sec)**
peanut butter				
molasses				
Murphy's Oil Soap™				
vegetable oil				
honey				

Name: _____   Date: _____

Analysis: Study the results of your experiment. Decide what the data means. This information can then be used to help you draw a conclusion about what you learned in your investigations.

Create a graph that will compare the average flow time of each liquid in the control group with the average flow time of each liquid in the experimental group. Place the dependent variable, time in seconds, on the y-axis. Place the independent variable, the type of liquid tested in the control group and experimental group, on the x-axis.

Affect of Temperature on Viscosity (flow of liquid)

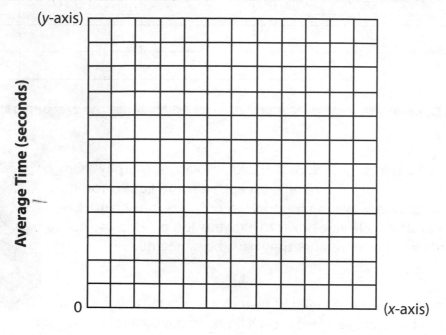

Type of Liquid in Control and Experimental Groups

Conclusion:
Write a summary of the experiment (what actually happened). It should include whether or not the hypothesis was supported by the data collected.

_____

_____

_____

_____

_____

Infer:
What can you infer about the temperature of magma and viscosity?

_____

_____

_____

Unit 8: Metamorphic Rocks
Teacher Information

Topic: Metamorphic rocks are the result of the process of physical forces in nature.

Standards:
    NSES Unifying Concepts and Processes, (D)
    NCTM Measurement
    STL Abilities for a Technological World
    See **National Standards** section (pages 61–65) for more information on each standard.

Concepts:
- Metamorphic rock is the rock formed when sedimentary or igneous rocks undergo a change due to pressure or heat in the earth.

Naïve Concepts:
- Rocks grow.
- All rocks are the same, and it's hard to tell how they originated.

Science Process Skills:

Students will **develop vocabulary** relating to metamorphic rocks. They will **identify** the main ingredients of metamorphic rocks. Students will **infer** the relationship between heat and pressure and metamorphic rock formations.

Lesson Planner:
1. <u>Directed Reading</u>: Introduce the concepts and essential vocabulary relating to metamorphic rocks using the directed reading exercise on the Student Information pages.
2. <u>Assessment</u>: Evaluate student comprehension of the information in the directed reading exercise using the quiz located on the Quick Check page.
3. <u>Concept Reinforcement</u>: Strengthen student understanding of concepts with the activities found on the Knowledge Builder page. **Materials Needed:** Activity #1—rolling pin, graham crackers, chocolate bars, mini-marshmallows, paper plates, resealable plastic bags, paper towels, honey, butterscotch chips, raisins, coconut, access to a microwave; Activity #2—large bowl, flour, salt, warm water, food coloring, paper plates, textbooks

Extension: Marble is one type of metamorphic rock that is used in building construction. Students research and compile a list of metamorphic rocks and their common uses.

Real World Application: The Appalachian Mountains, including the Blue Ridge Mountains, consist mainly of metamorphic rock.

Unit 8: Metamorphic Rocks
Student Information

Metamorphic rock is the type of rock formed when sedimentary rock or igneous rock undergoes a change due to pressure or heat in the earth. Marble is an example of a metamorphic rock. Limestone changes under heat and pressure, transforming into a new kind of rock, marble.

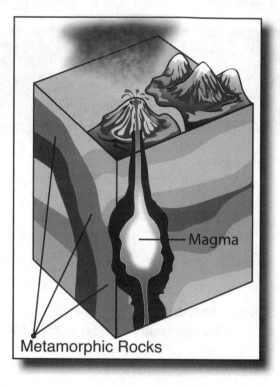

Metamorphism means "change" in rocks. Heat, pressure, and fluids that contain chemicals can convert mineral composition in rocks, thereby forming new, different rocks. Metamorphism can occur beneath the surface of the earth, where the magma from the asthenosphere comes in contact with the bottom of the crust of the earth. Here, heat is the greatest metamorphic agent. Tremendous amounts of heat and pressure can also be found along plate lines that are converging. Ultimately, metamorphism causes chemical changes in the composition of the rocks. Metamorphism can occur to varying degrees depending upon the levels/amounts of heat, pressure, and chemically active fluids.

There are two general classifications or families of metamorphic rocks: foliated and nonfoliated.

Foliated metamorphic rocks form with layered bands. Foliated rocks are sometimes confused with sedimentary rocks because of the layered bands. Common examples of foliated metamorphic rocks include slate, schist, and gneiss.

foliated meamorphic rock

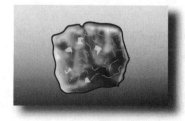

nonfoliated meamorphic rock

Nonfoliated metamorphic rocks form without layered bands. Common examples of nonfoliated metamorphic rocks include marble and quartzite.

Name: _____ Date: _____

Quick Check

Matching

_____ 1. metamorphism

_____ 2. foliated metamorphic rocks

_____ 3. marble

_____ 4. nonfoliated metamorphic rocks

_____ 5. gneiss

a. example of a nonfoliated metamorphic rock

b. example of a foliated metamorphic rock

c. "change"

d. form without layered bands

e. form with layered bands

Fill in the Blanks

6. _____, _____, and _____ that contain chemicals can convert mineral composition in rocks, thereby forming new, different rocks.

7. _____ _____ is the type of rock formed when sedimentary or igneous rocks undergo a change due to pressure or heat in the earth.

8. Metamorphism can occur beneath the surface of the earth, where the magma from the _____ comes in contact with the bottom of the _____ of the earth.

9. Ultimately, _____ causes chemical changes in the composition of the rocks.

10. Tremendous amounts of heat and pressure can also be found along plate lines that are _____.

Multiple Choice

11. Which of the following is NOT an example of foliated metamorphic rock?

    a. slate   b. schist

    c. gneiss   d. marble

12. Limestone changes under heat and pressure, transforming into a new kind of rock called _____.

    a. marble   b. slate

    c. gneiss   d. schist

13. How many general classifications or families of metamorphic rocks are there?

    a. three   b. six

    c. five   d. two

Name: _____ Date: _____

Knowledge Builder

Activity #1: Metamorphic S'more

Directions: Build a metamorphic rock. Use the ingredients below to represent materials within metamorphic rocks.

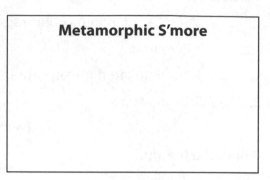

Metamorphic S'more

- graham crackers and chocolate bar: layers of rock
- mini-marshmallows, butterscotch chips, and raisins: smaller chunks of rock
- coconut: organic debris
- honey: chemically active liquid

Place one graham cracker half on a paper plate. Add a layer of mini-marshmallows, butterscotch chips, chocolate bar, coconut, honey, and raisins on top of the cracker. Place another graham cracker half on top of the last layer. Place the paper plate with s'more in a resealable plastic bag. Using a rolling pin, carefully apply pressure to the s'more. Take the paper plate with the smashed s'more from the bag. Place the paper plate and s'more in the microwave for 60 seconds. Record your observation in drawing form in the box above.

Conclusion: How is your metamorphic s'more similar to a metamorphic rock? _____

_____

_____

Activity #2: Play Dough Metamorphic Rock

Directions: Build a metamorphic rock. First make a batch of play dough. In a large bowl, place 1000 mL (1L) of flour and 250 mL of salt. Gradually add 250 mL of warm water to the mixture, and carefully mix together using your hands. Add more water as needed to make a doughy mixture. Divide dough into 4 balls and place in separate bowls. Add a different food coloring to each bowl and mix. When the mixture becomes a doughy substance, place it on a hard surface and knead it until it is smooth and rubbery. Next, use the following items to represent materials within metamorphic rocks. Make about two dozen different-colored pea-sized balls from the play dough. These represent rock particles. Place the play dough balls close together on a paper plate. Place a second paper plate on top of the play dough balls. Next, stack several text books on top of the paper plate. The books represent layers of rock building up on top of the rock particles. Heat builds as the rock particles are pushed deeper into the earth's crust. Remove the books and carefully peel away the paper plate. Examine the play dough metamorphic rock you made.

Conclusion: How was the formation of your play dough metamorphic rock similar to the formation of a metamorphic rock in nature? _____

_____

_____

Unit 9: The Rock Cycle
Teacher Information

Topic: The rock cycle is a diagram that is used to explain how Earth's processes change a rock from one type to another through geological time.

Standards:
>   **NSES** Unifying Concepts and Processes, (D)
>   **NCTM** Measurement
>   **STL** Abilities for a Technological World
>   See **National Standards** section (pages 61–65) for more information on each standard.

Concepts:
- Sedimentary, igneous, and metamorphic rock change in form and structure over and over again as a result of natural forces.

Naïve Concepts:
- Rocks grow.
- All rocks are the same, and it's hard to tell how they originated.

Science Process Skills:

Students will **develop vocabulary** relating to the rock cycle. They will **explain** how Earth's processes change a rock from one type to another through geological time. Students will **infer** the relationship between plate tectonics and the rock cycle.

Lesson Planner:
1. <u>Directed Reading</u>: Introduce the concepts and essential vocabulary relating to the rock cycle using the directed reading exercise found on the Student Information page.
2. <u>Assessment</u>: Evaluate student comprehension of the information in the directed reading exercise using the quiz located on the Quick Check page.
3. <u>Concept Reinforcement</u>: Strengthen student understanding of concepts with the activities found on the Knowledge Builder page. **Materials Needed:** Activity #1—aluminum foil, scissors, several different-colored crayons, plastic knife or crayon sharpener, clothespin, candle, matches, metric ruler, and rolling pin (**Caution:** close adult supervision is need for this activity.); Activity #2—poster board, variety of art supplies such as colored pencils, markers, glue, scissors, glitter

Extension: Students use modeling clay to create three-dimensional models of convergent, divergent, and transform boundaries.

Real World Application: Most of the earth's minerals, oil, and gas are located at a maximum depth of about three miles beneath the crust.

Unit 9: The Rock Cycle
Student Information

Rocks are formed in a cycle. **Igneous rocks** are formed as magma is extruded and cooled in hotspots, volcanoes, and in sea-floor spreading. When exposed rocks become subjected to elements, such as wind, rain, sun, freezing, thawing, glaciers, etc., they break into smaller parts and are often transported where they are deposited in layers. When the layers are subjected to tremendous pressure, **sedimentary rocks** are formed. Rocks may become **metamorphic** if the pressure is great enough and intense heat is present over time to produce a chemical change in the rocks. The rocks may even melt and become part of the magma once again. It is important to realize that rocks do not always go through all three phases of the rock cycle. For example, igneous rocks may become metamorphosed beneath the surface of the earth without ever being broken down into sediment.

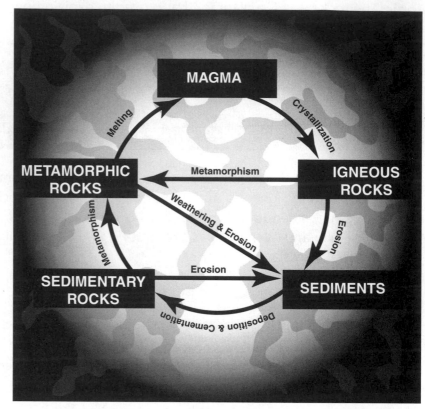

The **rock cycle** takes place both on continental and oceanic crust and includes the inner workings of the earth. Central to the rock cycle is the concept of plate tectonics. The **Theory of Plate Tectonics** explains how the continents were able to move to their present locations. Scientists believe the earth's outermost layer, the **lithosphere**, broke apart and formed seven large plates (pieces) and many smaller pieces. The motion of magma in the mantle, or **asthenosphere**, just under the crust, causes the movement of the plates. The plates move between 5 and 10 centimeters (cm) a year.

There are three types of plate boundaries: convergent boundaries, divergent boundaries, and transform boundaries. The plates sometimes collide and grind past each other at the plate boundaries. **Convergent plate boundaries** crash together, often forming mountains, like the Himalayas, and producing earthquake and volcanic activity. **Divergent plate boundaries** move apart. Most divergent boundaries are in the ocean. They build undersea mountain ranges called mid-ocean ridges as magma oozes out of the earth. **Transform boundaries** are where two plates slide past each other, often causing earthquakes. These plate movements are critical steps in introducing magma to the earth's surface. Equally important are convergent plate boundaries, whereby more dense oceanic crust slides under lighter, less dense continental crust. It is at these convergent plate boundaries that oceanic crust is literally recycled by virtue of tremendous heat and pressure, which ultimately results in melting as the crust reenters the magma of the asthenosphere.

Name: _____ Date: _____

Quick Check

Matching

_____ 1. metamorphic rock

_____ 2. Plate Tectonics

_____ 3. igneous rock

_____ 4. sedimentary rock

_____ 5. rock cycle

a. rocks that are formed by the crystallization of magma

b. rocks that are formed by the layering of sediments

c. takes place both on continental and oceanic crust and includes the inner workings of the earth

d. formed by chemical alteration of previously existing rocks, due to heat and pressure

e. theory used to explain how the continents were able to move to their present locations

Fill in the Blanks

6. Scientists believe the earth's outermost layer, the _____, broke apart and formed seven large plates (pieces).

7. _____ _____ are formed as magma is extruded and cooled in hotspots, volcanoes, and in sea-floor spreading.

8. Igneous rocks may become _____ beneath the surface of the earth without ever being broken down into sediment.

9. It is important to realize that rocks do not always go through all three _____ of the rock cycle.

10. The motion of magma in the mantle, or _____, just under the crust, causes the movement of the plates.

Multiple Choice

11. Earth's plates move between _____ a year.

   a. 5 and 10 cm
   b. 10 and 15 cm
   c. 20 and 25 cm
   d. 30 and 35 cm

12. How many types of plate boundaries are there?

   a. four
   b. three
   c. seven
   d. two

13. If the pressure is great enough and intense heat is present over time to produce a chemical change in the rocks, rocks may become _____.

   a. sedimentary
   b. metamorphic
   c. igneous
   d. volcanoes

Name: _____   Date: _____

Knowledge Builder

Activity #1: Rock Model

Directions: Cut two pieces of foil 30 cm X 20 cm. Place the two pieces of foil on top of each other, forming a double layer. Remove the paper from several different-colored crayons. Using one crayon, make a pile of "sediment" by sharpening the crayon or scraping a plastic knife up and down the length of the crayon. Shave or scrape the other crayons until you have a pile of shavings that measures about 8 cm X 8 cm and approximately 6 cm thick. Put the shavings on one half of the doubled foil, fold the foil over the shavings, and bend the edges up to prevent

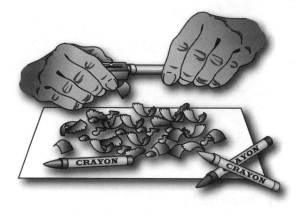

spilling. Flatten the shavings with a rolling pin. Roll the edges so no wax can drip out, and attach a clothespin to one side. Hold the foil over a candle flame for several minutes. (Caution: adult supervision is needed for this part of the activity.) Let cool. Unwrap and examine.

Conclusion: What kind of rock formed when you added heat? _____

_____

_____

Activity #2: Rock Cycle Poster

Directions: Create a poster of the rock cycle.

- Describe and illustrate the rock cycle.
- Describe the relationship between the three rock types: igneous, metamorphic, and sedimentary.
- Explain how igneous, metamorphic, and sedimentary rocks are formed.
- Explain how each type of rock is classified.
- Identify where igneous, metamorphic, and sedimentary rocks are most likely to be found.
- Include pictures of examples of each type of rock.

Unit 10: Sand
Teacher Information

Topic: Sand is formed from the weathering of rocks and minerals.

Standards:
NSES Unifying Concepts and Processes, (D)
NCTM Measurement
STL Abilities for a Technological World
See **National Standards** section (pages 61–65) for more information on each standard.

Concepts:
- Sand is formed from the weathering of rocks and minerals.
- Sand is transported by wind and water and deposited in the form of beaches, dunes, sand spits, and sand bars.

Naïve Concepts:
- Sand is made from glass.
- Sand rises to the surface from the ground below it.

Science Process Skills:
Students will **develop vocabulary** relating to sand. Students will **infer** processes that shape sand and its location. Students will **identify** conditions that result in various types of sand. Students will **associate** sand with water.

Lesson Planner:
1. Directed Reading: Introduce the concepts and essential vocabulary relating to sand formation using the directed reading exercise found on the Student Information page.
2. Assessment: Evaluate student comprehension of the information in the directed reading exercise using the quiz located on the Quick Check page.
3. Concept Reinforcement: Strengthen student understanding of concepts with the activities found on the Knowledge Builder page. **Materials Needed:** Activity #1—different types of sand (beach sand, sand from a lake or stream, sand for a sand box), toothpicks, black paper, hand lenses, magnet; Activity #2—electric fan, box, sand, pencil

Extension: Students investigate the relationship between sand beaches and tides.

Real World Application: White Sands National Monument is a unique dune field located in New Mexico. White glistening sand dunes of gypsum cover 275 miles of desert.

Unit 10: Sand
Student Information

Sand consists of small rock and mineral particles ranging in size from 0.06 to 2 millimeters (mm) in diameter. Sand contains large amounts of quartz; lime, gypsum, feldspar, and iron ore may also be found in sand. Most grains of sand were once a part of solid rock. The rocks break down over time due to chemical reactions with air and water or the crashing of waves against a coastline.

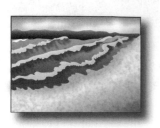

Sand creates coastal landforms including beaches, sand dunes, sandbars, and sand spits. A **beach** is an area of sand that slopes down to the water's edge of a sea or lake. Sand is transported to beaches by the movement of water and wave energy. Beaches are formed where water moves more sand particles toward the shore than away from it. Beaches are representative of the rocks native to the area.

Rivers are another common source of shoreline sand. As with beaches, sand is transported by water and deposited along the shore. Sand from rivers typically originates from rocks and minerals within the river valley.

Mounds of wind-blown sand are called dunes. **Sand dunes** are simply piles of sand. Dunes are formed when wind and waves transport sand onto the beach. They are present on shorelines where fine sediment is transported landward by a combination of wind and waves. Sand dunes are also found in deserts. Large amounts of sand that wash up from the shallow sea bottoms can form small sand dunes.

Sandbars are long, narrow ridges built from an accumulation of sand along the outer part of the shore. They are formed by the motion of waves and tides. Sandbars are affected by the action of the water. They change constantly depending on the forces put on them by tidal or wave action.

A **sand spit** is a linear deposition of sand that forms off coastlines. Spits connect to land at one end and look like a finger-like piece of land that extends into the open water. Wind and water carry enough sand and gravel to create these narrow features.

Name: _____ Date: _____

Quick Check

Matching

_____ 1.   sandbars

_____ 2.   sand spits

_____ 3.   sand

_____ 4.   dunes

_____ 5.   beach

a.   consists of small rock and mineral particles ranging in size from 0.06 to 2 mm in diameter

b.   piles of sand on beaches or in deserts

c.   long, narrow ridges built from an accumulation of sand along the outer part of the shore

d.   area of sand that slopes down to the water's edge of a sea or lake

e.   linear deposition of sand that forms off coastlines

Fill in the Blanks

6.   Sand is transported to beaches by the movement of _____ and _____ energy.

7.   Sand from rivers typically originates from _____ and _____ within the river valley.

8.   Sandbars are affected by the action of the _____.

9.   _____ connect to land at one end and look like a finger-like piece of land that extends into the open water.

10. _____ are formed where water moves more sand particles toward the shore than away from it.

Multiple Choice

11. Mounds of wind-blown sand are called _____.
   a.   beaches   b.   sand spits
   c.   sandbars   d.   sand dunes

12. Sand consists of small rock and mineral particles ranging in size from _____.
   a.   0.06 to 2 mm   b.   0.60 to 20 mm
   c.   0.16 to 2 mm   d.   0.63 to 23 mm

13. Which of the following landforms is not created from sand?
   a.   beaches   b.   volcanoes
   c.   sand spits   d.   sandbars

Name: _____ Date: _____

Knowledge Builder

Activity #1: Minerals Found in Sand

Directions: Pour a sample of sand onto a black piece of paper. With a toothpick, separate the grains. Notice the different shapes, sizes, and colors of the sand. Use a hand lens to examine the sand. Record your observations in the data table below. Repeat with each type of sand.

Minerals	Description	Mineral Present (Yes/No)		
		Sand #1	Sand #2	Sand #3
Quartz	Grains are glassy and colorless crystals.			
Garnet	Crystals are red.			
Mica	Crystals are thin, flaky, and black.			
Hornblende	Crystals are rectangular and black.			
Magnetite	Crystals are black and attracted to a magnet.			
Amethyst	Crystals are purple.			
Olivine	Crystals are green.			

Conclusion: What is sand? _____

_____

_____

_____

Activity #2: Making Sand Dunes

Directions: Sand dunes act as a buffer zone, protecting the land behind them from the force of ocean wind and waves. Pour a pile of dry sand on the bottom of a large box cut away as shown in the diagram. Place an electric fan facing the box. Turn the fan on low. Observe how the particles are moved. Next, turn the fan on high.

Observation

1. What happened when the velocity of the wind was increased?

    _____

2. Put a pencil in the path of the blowing sand. What happened?

    _____

    _____

Conclusion: How is the build up of sand in this activity similar to the formation of sand dunes?

_____

_____

_____

Unit 11: Fossils
Teacher Information

Topic: Fossils provide clues to the past animal and plant life on Earth.

Standards:
   NSES Unifying Concepts and Processes, (D)
   NCTM Geometry and Measurement
   STL Abilities for a Technological World
   See **National Standards** section (pages 61–65) for more information on each standard.

Concepts:

- Fossils provide clues to the past plant and animal life on Earth.
- Fossils provide clues to the order in which layers of rocks were formed.

Naïve Concepts:

- Fossils are pieces of dead animals and plants.
- Fossils only represent bones and shells of extinct animals.

Science Process Skills:

Students will **develop vocabulary** relating to fossils. Students will **infer** that fossils are clues to the past plant and animal life on Earth. Students will **create** model molds of modern fossils.

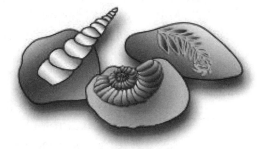

Lesson Planner:

1. Directed Reading: Introduce the concepts and essential vocabulary relating to fossils using the directed reading exercise found on the Student Information page.
2. Assessment: Evaluate student comprehension of the information in the directed reading exercise using the quiz located on the Quick Check page.
3. Concept Reinforcement: Strengthen student understanding of concepts with the activities found on the Knowledge Builder page. **Materials Needed:** Activity #1—large bowl, flour, salt, warm water, waxed paper, rolling pin, meter ruler, shells, leaves, twigs, fern fronds, cookie sheet, oven; Activity #2—plaster of Paris, bowl, vegetable oil or petroleum jelly, clean chicken bone, rubber bands, liquid latex

Extension: Students analyze local fossils for clues to living conditions in the past for their area.

Real World Application: The Petrified Forest National Park in Arizona has the greatest concentration of wood that has turned into stone. This type of fossil is called petrified wood. Logs of petrified wood (some as long as 200 feet) have been found in the park.

Unit 11: Fossils
Student Information

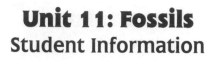

Fossils are the preserved remains, impressions, or traces of animals, plants, and other organisms. Fossils can be formed in several ways. Occasionally, **entire organisms** are preserved in some type of material. Insects are sometimes preserved in resins from trees, which eventually becomes

amber. Giant woolly mammoths have been discovered in the frozen tundra of the north. More typical are fossil remains of the hard parts of animals. Fossils are preserved mainly in sedimentary rock. The skeletal remains (shells, bones, teeth) of living organisms are surrounded by sediment that settles on the bottom of a body of water, and the remains become trapped when the sediments harden into rock.

Fossils provide clues to the types of plants and animals that were present many years ago. Fossils are also important clues in determining the order in which layers of rocks were formed. Since scientists know how old certain fossils are, when they find these fossils in a layer of rock, they can tell when the layer was formed. Such fossils are called **index fossils**.

Fossils may form in a variety of ways:

1. Some creatures have fallen or gotten stuck in tar—natural asphalt like that found in California's LaBrea Tar Pits. Insects often get stuck in tree sap that later hardens to amber. Sometimes animals may be freeze-dried in a glacier or on a mountaintop, like the 5,000-year old "ice man" found in the Italian Alps.

2. When creatures or plants get buried in sediments and later soak in mineral-laden water, they may literally turn to stone and become petrified. This kind of preservation may show details down to the level of individual cells, like Arizona's Petrified Forest.

3. Animals or plants buried in lake sediments or swamps that slowly decay while being squeezed and slow-cooked will become carbonized.

4. Sometimes only a creature's shape is preserved as either a cast or mold. Eruptions of Mt. Vesuvius in Italy have covered humans with volcanic ash. The ash hardens into a mold, while the body inside decays. If the mold later fills with sediment, a cast forms that reproduces the original body shape.

5. Tunnels or footprints left in moist sand will sometimes dry quickly, be buried, and then be compressed into stone. Fossil dung may get covered quickly and later petrify.

Fossils provide clues to 99% of Earth's organisms that have become **extinct** (no longer exist). **Paleontologists** study fossils to reconstruct Earth's living past.

Types of Fossils

Type of Fossil	Description	Example
Mold Fossils	Impressions are left by plants and animals in a rock after the plants or animals have decayed.	
Cast Fossils	Minerals collect in the mold of what was the plant or animal, forming a cast, or model of the original organism.	
Petrified fossils	Minerals replace the hard parts (teeth, bone, shell) of animals or plants, turning them into rock.	
Coprolites	Animal dung is petrified or turned into rock.	
Imprints	Impressions of parts of organisms are left in sediment before the sediment hardens.	

Name: _____   Date: _____

Quick Check

Matching

_____ 1. fossils

_____ 2. index fossils

_____ 3. coprolites

_____ 4. extinct

_____ 5. paleontologists

a. no longer exist

b. study fossils to reconstruct Earth's living past

c. fossils that indicate the order in which layers of rocks were formed

d. preserved remains, impressions, or traces of animals, plants, and other organisms

e. petrified animal dung

Fill in the Blanks

6. Insects are sometimes preserved in resins from trees, which eventually become _____.

7. Giant woolly _____ have been discovered in the frozen tundra of the north.

8. _____ are preserved mainly in sedimentary rock.

9. Fossils provide clues to the types of _____ and _____ that were present many years ago.

10. When creatures or plants get buried in sediments and later soak in mineral-laden water, they may literally turn to stone and become _____.

Multiple Choice

11. Impressions left by plants and animals in a rock after the plants or animals have decayed.
    a. coprolites
    c. mold fossils
    b. petrified fossils
    d. cast fossils

12. Minerals replace the hard parts (teeth, bone, shell) of animals or plants, turning them into rock.
    a. cast fossils
    c. coprolites
    b. petrified fossils
    d. mold fossils

13. Animal dung becomes petrified.
    a. petrified fossils
    c. mold fossils
    b. cast fossils
    d. coprolites

Name: _____ Date: _____

Knowledge Builder

Activity #1: Fossil Imprint

Directions: Make a batch of play dough. In a large bowl, place 1000 mL (1 L) of flour and 250 mL of salt. Gradually add 250 mL of warm water and carefully mix together using your hands. Add more water as needed until it becomes a doughy substance. Then place it on a hard surface and knead it until it is smooth and rubbery. Take a handful of dough and place it on a sheet of wax paper. Using a rolling pin, roll the dough to 1 cm thickness. Choose an item to be imprinted such as a shell, leaf, twig, or fern frond. Place a sheet of waxed paper over the objects and gently roll over the top of the objects with a rolling pin. Remove the waxed paper and items used for imprinting. Place the mold on a cookie sheet. Bake at 250 degrees in an oven until hard.

Conclusion: What information could a scientist infer about the past from your fossil? _____

_____

_____

Activity #2: Fossil Cast

Directions: Mix plaster of Paris to the consistency of pancake batter. Pour a thin layer into a small, shallow pan. Coat an object such as a chicken bone with a thin layer of petroleum jelly; lightly push the object into the surface of the plaster of Paris and allow it to dry. Coat the entire surface with a thin layer of petroleum jelly, including the bone. Pour another layer of plaster of Paris over the top. When the mold is dry, split it apart at the seam and remove the bone. On one half of the plaster, bore a thin channel from the edge to the impression left by the bone. Reattach both halves of the plaster of Paris and secure them tightly with large rubber bands. Pour liquid latex into the channel until the hole left by the bone is full. Allow to dry completely. Reopen the halves, and discover the cast of the bone.

Conclusion: What information could a scientist infer about the past from your cast? _____

_____

_____

Name: _____ Date: _____

Inquiry Investigation Rubric

Category	4	3	2	1
Participation	Used time well, cooperative, shared responsibilities, and focused on the task.	Participated, stayed focused on task most of the time.	Participated, but did not appear very interested. Focus was lost on several occasions.	Participation was minimal OR student was unable to focus on the task.
Components of Investigation	All required elements of the investigation were correctly completed and turned in on time.	All required elements were completed and turned in on time.	One required element was missing/or not completed correctly.	The work was turned in late and/or several required elements were missing and/or completed incorrectly.
Procedure	Steps listed in the procedure were accurately followed.	Steps listed in the procedure were followed.	Steps in the procedure were followed with some difficulty.	Unable to follow the steps in the procedure without assistance.
Mechanics	Flawless spelling, punctuation, and capitalization.	Few errors.	Careless or distracting errors.	Many errors.

Comments:

National Standards in Science, Math, and Technology

NSES Content Standards (NRC, 1996)

National Research Council (1996). *National Science Education Standards.* Washington, D.C.: National Academy Press.

UNIFYING CONCEPTS: K-12

Systems, Order, and Organization: The natural and designed world is complex. Scientists and students learn to define small portions for the convenience of investigation. The units of investigation can be referred to as systems. A system is an organized group of related objects or components that form a whole. Systems can consist of machines.

Systems, Order, and Organization

The goal of this standard is to ...

- Think and analyze in terms of systems.
- Assume that the behavior of the universe is not capricious. Nature is predictable.
- Understand the regularities in a system.
- Understand that prediction is the use of knowledge to identify and explain observations.
- Understand that the behavior of matter, objects, organisms, or events has order and can be described statistically.

Evidence, Models, and Explanation

The goal of this standard is to ...

- Recognize that evidence consists of observations and data on which to base scientific explanations.
- Recognize that models have explanatory power.
- Recognize that scientific explanations incorporate existing scientific knowledge (laws, principles, theories, paradigms, models), and new evidence from observations, experiments, or models.
- Recognize that scientific explanations should reflect a rich scientific knowledge base, evidence of logic, higher levels of analysis, greater tolerance of criticism and uncertainty, and a clear demonstration of the relationship between logic, evidence, and current knowledge.

Change, Constancy, and Measurement

The goal of this standard is to ...

- Recognize that some properties of objects are characterized by constancy, including the speed of light, the charge of an electron, and the total mass plus energy of the universe.
- Recognize that changes might occur in the properties of materials, position of objects, motion, and form and function of systems.
- Recognize that changes in systems can be quantified.
- Recognize that measurement systems may be used to clarify observations.

National Standards in Science, Math, and Technology (cont.)

Form and Function

The goal of this standard is to …
- Recognize that the form of an object is frequently related to its use, operation, or function.
- Recognize that function frequently relies on form.
- Recognize that form and function apply to different levels of organization.
- Enable students to explain function by referring to form, and explain form by referring to function.

NSES Content Standard A: Inquiry
- Abilities necessary to do scientific inquiry
  - Identify questions that can be answered through scientific investigations.
  - Design and conduct a scientific investigation.
  - Use appropriate tools and techniques to gather, analyze, and interpret data.
  - Develop descriptions, explanations, predictions, and models using evidence.
  - Think critically and logically to make relationships between evidence and explanations.
  - Recognize and analyze alternative explanations and predictions.
  - Communicate scientific procedures and explanations.
  - Use mathematics in all aspects of scientific inquiry.
- Understanding about inquiry
  - Different kinds of questions suggest different kinds of scientific investigations.
  - Current scientific knowledge and understanding guide scientific investigations.
  - Mathematics is important in all aspects of scientific inquiry.
  - Technology used to gather data enhances accuracy and allows scientists to analyze and quantify results of investigations.
  - Scientific explanations emphasize evidence, have logically consistent arguments, and use scientific principles, models, and theories.
  - Science advances through legitimate skepticism.
  - Scientific investigations sometimes result in new ideas and phenomena for study, generate new methods or procedures, or develop new technologies to improve data collection.

NSES Content Standard B: Properties and Changes of Properties in Matter 5-8

NSES Content Standard D: Structure of the Earth System 5-8

NSES Content Standard D: Earth in the Solar System 5-8

NSES Content Standard E: Science and Technology 5-8
- Abilities of technological design
  - * Identify appropriate problems for technological design.
  - * Design a solution or product.

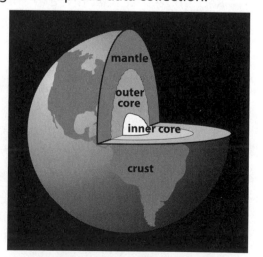

National Standards in Science, Math, and Technology (cont.)

* Implement the proposed design.
* Evaluate completed technological designs or products.
* Communicate the process of technological design.
- Understanding about science and technology
  * Scientific inquiry and technological design have similarities and differences.
  * Many people in different cultures have made and continue to make contributions.
  * Science and technology are reciprocal.
  * Perfectly designed solutions do not exist.
  * Technological designs have constraints.
  * Technological solutions have intended benefits and unintended consequences.

NSES Content Standard F: Science in Personal and Social Perspectives 5-8
- Science and technology in society
  * Science influences society through its knowledge and world view.
  * Societal challenges often inspire questions for scientific research.
  * Technology influences society through its products and processes.
  * Scientists and engineers work in many different settings.
  * Science cannot answer all questions, and technology cannot solve all human problems.

NSES Content Standard G: History and Nature of Science 5-8
- Science as human endeavor
- Nature of science
  * Scientists formulate and test their explanations of nature using observation, experiments, and theoretical and mathematical models.
  * It is normal for scientists to differ with one another about interpretation of evidence and theory.
  * It is part of scientific inquiry for scientists to evaluate the results of other scientists' work.
- History of science
  * Many individuals have contributed to the traditions of science.
  * Science has been and is practiced by different individuals in different cultures.
  * Tracing the history of science can show how difficult it was for scientific innovators to break through the accepted ideas of their time to reach the conclusions we now accept.

National Standards in Science, Math, and Technology (cont.)

Standards for Technological Literacy (STL) ITEA, 2000

International Technology Education Association (2000). *Standards for Technological Literacy.* Reston, VA: International Technology Education Association.

The Nature of Technology

Students will develop an understanding of the:

1. Characteristics and scope of technology.
2. Core concepts of technology.
3. Relationships among technologies and the connections between technology and other fields of study.

Technology and Society

Students will develop an understanding of the:

4. Cultural, social, economic, and political effects of technology.
5. Effects of technology on the environment.
6. Role of society in the development and use of technology.
7. Influence of technology on history.

Design

Students will develop an understanding of the:

8. Attributes of design.
9. Engineering design.
10. Role of troubleshooting, research and development, invention and innovation, and experimentation in problem solving.

Abilities for a Technological World

Students will develop abilities to:

11. Apply the design process.
12. Use and maintain technological products and systems.
13. Assess the impact of products and systems.

The Designed World

Students will develop an understanding of and be able to select and use:

14. Medical technologies.
15. Agricultural and related biotechnologies.
16. Energy and power technologies.
17. Information and communication technologies.
18. Transportation technologies.
19. Manufacturing technologies.
20. Construction technologies.

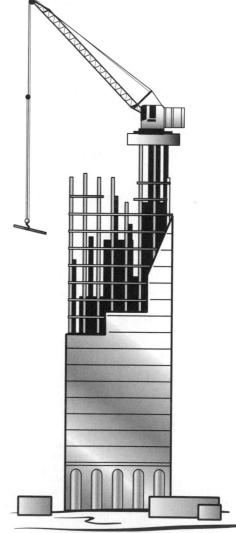

National Standards in Science, Math, and Technology (cont.)

Principles and Standards for School Mathematics (NCTM), 2000

National Council for Teachers of Mathematics (2000). *Principles and Standards for School Mathematics.*
Reston, VA: National Council for Teachers of Mathematics.

Number and Operations
Students will be enabled to:
- Understand numbers, ways of representing numbers, relationships among numbers, and number systems.
- Understand meanings of operations and how they relate to one another.
- Compute fluently and make reasonable estimates.

Algebra
Students will be enabled to:
- Understand patterns, relations, and functions.
- Represent and analyze mathematical situations and structures using algebraic symbols.
- Use mathematical models to represent and understand quantitative relationships.
- Analyze change in various contexts.

Geometry
Students will be enabled to:
- Analyze characteristics and properties of two- and three-dimensional geometric shapes and develop mathematical arguments about geometric relationships.
- Specify locations and describe spatial relationships using coordinate geometry and other representational systems.
- Apply transformations and use symmetry to analyze mathematical situations.
- Use visualization, spatial reasoning, and geometric modeling to solve problems.

Measurement
Students will be enabled to:
- Understand measurable attributes of objects and the units, systems, and processes of measurement.
- Apply appropriate techniques, tools, and formulas to determine measurements.

Data Analysis and Probability
Students will be enabled to:
- Formulate questions that can be addressed with data and collect, organize, and display relevant data to answer them.
- Select and use appropriate statistical methods to analyze data.
- Develop and evaluate inferences and predictions that are based on data.
- Understand and apply basic concepts of probability.

Science Process Skills

Introduction: Science is organized curiosity, and an important part of this organization includes the thinking skills or information-processing skills. We ask the question "why?" and then must plan a strategy for answering the question or questions. In the process of answering our questions, we make and carefully record observations, make predictions, identify and control variables, measure, make inferences, and communicate our findings. Additional skills may be called upon, depending on the nature of our questions. In this way, science is a verb, involving active manipulation of materials and careful thinking. Science is dependent on language, math, and reading skills, as well as the specialized thinking skills associated with identifying and solving problems.

BASIC PROCESS SKILLS:

Classifying: Grouping, ordering, arranging, or distributing objects, events, or information into categories based on properties or criteria, according to some method or system.

> Example: Classifying rocks and minerals, e.g., three major classifications of rocks and their families.

Observing: Using the senses (or extensions of the senses) to gather information about an object or event.

> Example: Observing physical properties of rocks and minerals; critical for compare and contrast skills, which are needed for analysis.

Measuring: Using both standard and nonstandard measures or estimates to describe the dimensions of an object or event. Making quantitative observations.

> Example: Measuring the density of minerals.

Inferring: Making an interpretation or conclusion based on reasoning to explain an observation.

> Example: Stating the types of life and living conditions from fossil evidence.

Communicating: Communicating ideas through speaking or writing. Students may share the results of investigations, collaborate on solving problems, and gather and interpret data both orally and in writing. Using graphs, charts, and diagrams to describe data.

> Example: Describing an event or a set of observations. Participating in brainstorming and hypothesizing before an investigation. Formulating initial and follow-up questions in the study of a topic. Summarizing data, interpreting findings, and offering conclusions. Questioning or refuting previous findings.

Science Process Skills (cont.)

Predicting: Making a forecast of future events or conditions in the context of previous observations and experiences.

> Example: Stating, "Magma containing high levels of silica will cause volcanoes to explode more violently."

Manipulating Materials: Handling or treating materials and equipment skillfully and effectively.

> Example: Using a rock hammer, hand lens, and sieves to analyze rock samples.

Using Numbers: Applying mathematical rules or formulas to calculate quantities or determine relationships from basic measurements.

> Example: Determining the relative thickness of the layers of the earth.

Developing Vocabulary: Specialized terminology and unique uses of common words in relation to a given topic need to be identified and given meaning.

> Example: Using context clues, working definitions, glossaries or dictionaries, word structure (roots, prefixes, suffixes), and synonyms and antonyms to clarify meaning, i.e., sedimentary, metamorphic, and igneous rocks; minerals; Mohs' scale; ore; mantle; core; lithosphere; plate tectonics; convergent plate boundaries; divergent plate boundaries.

Questioning: Questions serve to focus inquiry, determine prior knowledge, and establish purposes or expectations for an investigation. An active search for information is promoted when questions are used.

> Example: Using what is already known about a topic or concept to formulate questions for further investigation; hypothesizing and predicting prior to gathering data; or formulating questions as new information is acquired.

Using Clues: Key words and symbols convey significant meaning in messages. Organizational patterns facilitate comprehension of major ideas. Graphic features clarify textual information.

> Example: Listing or underlining words and phrases that carry the most important details, or relating key words together to express a main idea or concept.

Science Process Skills (cont.)

INTEGRATED PROCESS SKILLS

Creating Models: Displaying information by means of graphic illustrations or other multisensory representations.

> Example: Drawing a graph or diagram; constructing a three-dimensional object, e.g., Earth model; using a digital camera to record an event; constructing a chart or table; or producing a picture or map that illustrates information about the earth and its geologic features.

Formulating Hypotheses: Stating or constructing a statement that is testable about what is thought to be the expected outcome of an experiment (based on reasoning).

> Example: Making a statement to be used as the basis for an experiment: "If the substance is a combination of minerals, it must be some sort of rock."

Generalizing: Drawing general conclusions from particulars.

> Example: Making a summary statement following analysis of experimental results: "The fossil evidence and layers of this rock indicate that it is some sort of sedimentary rock."

Identifying and Controlling Variables: Recognizing the characteristics of objects or factors in events that are constant or change under different conditions and that can affect an experimental outcome, keeping most variables constant while manipulating only one variable.

> Example: Controlling variables in an experiment to determine the viscosity of substances representing magmas.

Defining Operationally: Stating how to measure a variable in an experiment; defining a variable according to the actions or operations to be performed on or with it.

> Example: Defining minerals as solid, inorganic crystalline substances with definite chemical compositions.

Recording and Interpreting Data: Collecting bits of information about objects and events that illustrate a specific situation, organizing and analyzing data that has been obtained, and drawing conclusions from it by determining apparent patterns or relationships in the data.

> Example: Recording data (taking notes, making lists/outlines, recording numbers on charts/graphs, making tape recordings, taking photographs, writing numbers of results of observations/measurements) from observations to determine the physical properties of rocks and minerals.

Science Process Skills (cont.)

Making Decisions: Identifying alternatives and choosing a course of action from among alternatives after basing the judgment for the selection on justifiable reasons.

> Example: Determining optimum location(s) for collecting rocks and minerals.

Experimenting: Being able to conduct an experiment, including asking an appropriate question, stating a hypothesis, identifying and controlling variables, operationally defining those variables, designing a "fair" experiment, and interpreting the results of an experiment.

> Example: Formulating a researchable question, identifying and controlling variables including a manipulated and responding variable, data collection, data analysis, drawing conclusions, and formulating new questions as a result of the conclusions.

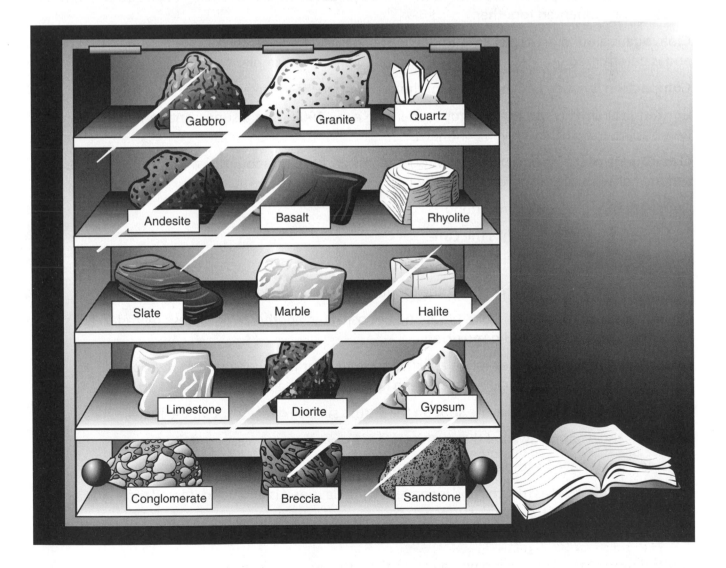

Definitions of Terms

Insects are sometimes preserved in resins from trees, which eventually become **amber**.

Almost plastic-like in rigidity, the upper mantle is known as the **asthenosphere**.

A **beach** is an area of sand that slopes down to the water's edge of a sea or lake.

Cast fossils are formed by minerals that have collected in the mold of what was the plant or animal, forming a cast or model of the original organism.

Cementation is a chemical process through which water carries and deposits dissolved minerals in the small spaces between the sediment particles.

Chemical sedimentary rocks form from minerals left behind after water evaporates.

Chemical weathering causes changes in the chemical makeup of rocks and makes them crumble. Chemical weathering in caves causes stalactites and stalagmites.

Clastic sedimentary rocks are formed when little pieces of broken-up rock that pile up over time are pressed and cemented together.

Cleavage is a mineral test relating to the crystalline structure and chemical bonding of the mineral and is determined by close observation of breaks within the sample.

Compaction is caused by the weight of the overlying rocks squeezing the sediments together.

Continental Drift is the theory proposed in the early 1900s that a supercontinent called Pangaea broke apart, forming the seven continents.

Convergent plate boundaries crash together, often forming mountains, like the Himalayas, and producing earthquake and volcanic activity.

Coprolites are petrified animal dung.

The innermost layer of Earth is the **core**.

The outermost layer of relatively thin **crust** is also called the lithosphere.

Sediments from rivers form **deltas** at the mouths of rivers before entering the ocean.

Erosion takes away the soil in one place and deposits it in another. When these materials are swept to a new location, it is called **deposition**.

Divergent plate boundaries move apart. Most divergent boundaries are in the ocean. They build undersea mountain ranges called mid-ocean ridges as magma oozes out of the earth.

The earthquake's **epicenter** is the spot on the earth's surface directly above the focus.

Erosion is the wearing away of the earth's surface by wind, water, ice, or gravity.

Organisms that have become **extinct** no longer exist.

Molten rock is known as magma and is converted into lava as it is **extruded** (forced through an opening in the earth) onto the earth's surface to cool.

Extrusive igneous rocks are formed when lava cools on the earth's surface.

Faulted mountains are formed when the earth's crust breaks into huge blocks.

Definitions of Terms (cont.)

Felics are igneous rocks that are high in silica content (+65%); lightweight; main minerals include orthoclastic feldspar, mica, and quartz; rock examples include rhyolite and granites.

The **focus** is the place in the earth's crust where the pressure was released during an earthquake.

Folded mountains are formed when the earth's crust folds into great waves.

Foliated metamorphic rocks form with layered bands.

Fossils are the preserved remains, impressions, or traces of animals, plants, and other organisms.

Geology is the study of the earth.

Hardness is a test for how hard a mineral is.

Soil forms distinct layers known as **horizons**.

Igneous rock, known as "fire rock," is rock formed by the cooling of melted material such as magma inside the earth and lava above the ground.

Lava is cooled magma on the surface of the earth.

Imprints are impressions of parts of organisms left in sediment before it hardens.

Fossils are important clues in determining the order in which layers of rocks were formed. Such fossils are called **index fossils**.

Intermediates are igneous rocks that contain 55–65% silica; medium color; main minerals include hornblende, biotite mica, and pyroxene; rock examples include andesite and diorite.

Intrusive igneous rocks are formed when magma cools inside the earth.

The outermost layer of relatively thin crust is also called the **lithosphere**.

Luster is a test for how light interacts with the surface of the mineral.

Mafics are igneous rocks that are high in iron and magnesium; lower in silica content (45–55%); usually darker in color; main minerals include plagioclastic feldspar, olivine, and pyroxene; rock examples include basalts and gabbros.

Molten rock is known as **magma** and is converted into lava as it is extruded (forced through an opening in the earth) onto the earth's surface to cool.

Magnetism is when minerals contain magnetic properties; this is easily tested by placing the mineral near a magnet.

Directly below the lithosphere is the **mantle**, a layer of dense, molten rock about 2,970 km (1,856 miles) thick.

Mechanical weathering is the breaking of rock into smaller pieces by physical forces such as wind, water, plants, and freezing.

Metamorphic rock is the rock formed when sedimentary or igneous rocks undergo a change due to pressure or heat in the earth.

Metamorphism means "change" in rocks.

Definitions of Terms (cont.)

A **mineral** is a naturally occurring inorganic (nonliving) solid. It has a crystalline structure.

Mohs' scale is an index of minerals of varying hardness.

Mold fossils are impressions left by plants and animals in a rock after the plants or animals have decayed.

Nonfoliated metamorphic rocks form without layered bands.

Organic sedimentary rocks, such as coal and organic limestone, contain the remains of materials that were once alive.

Paleontologists study fossils to reconstruct Earth's living past.

Scientists believe the seven continents were once part of a supercontinent called **Pangaea**.

Today, scientists use the Theory of **Plate Tectonics** to explain how the continents were able to move to their present locations.

Reaction to acid is a mineral test that shows a reaction (usually fizzing/bubbling). These minerals tend to come from the carbonate family.

The **rock cycle** takes place both on continental and oceanic crust and includes the inner workings of the earth. Central to the rock cycle is the concept of plate tectonics.

Rocks are mixtures that are usually made up of an assortment of minerals.

Sand consists of small rock and mineral particles ranging in size from 0.06 to 2 mm in diameter.

Sandbars are long, narrow ridges built from an accumulation of sand along the outer part of the shore.

Sand dunes are simply piles of sand.

A **sand spit** is a linear deposition of sand that forms off coastlines.

Sediment is small pieces of rock, sand, clay, and other materials picked up by flowing water.

Sedimentary rock is a kind of rock formed when a layer of sediment becomes solid.

Silicates are combinations of other chemical elements with silicon and oxygen.

Silicates have a common crystalline structure, known as the **silicon-oxygen tetrahedron**.

Soil is a mixture of nonliving things, such as sand grains, smaller rock particles, and minerals.

Specific Gravity is a mineral test that refers to the weight of an amount of mineral compared to an equal amount of water.

Strata are layers of rock.

Streak color is the color a mineral leaves when dragged across a piece of unglazed white porcelain tile.

Transform boundaries form where two plates slide past each other.

Volcanic mountains are formed when converging boundaries crash together.

Weathering is a change in the physical and/or chemical composition of a rock due to the forces of nature.

Answer Keys

Historical Perspective
Quick Check (page 8)
Matching
1. e   2. c   3. a   4. d   5. b

Fill in the Blanks
6. Bertram Boltwood
7. Friedrich Mohs
8. Leonardo da Vinci
9. *Principles of Geology*
10. Harry Hess

Multiple Choice
11. b   12. c   13. a

Layers of the Earth
Quick Check (page 12)
Matching
1. c   2. d   3. a   4. b   5. e

Fill in the Blanks
6. rock
7. mantle
8. outer core, inner core
9. Plate Tectonics
10. seismic waves

Multiple Choice
11. d   12. c   13. a

Rocks and Minerals
Quick Check (page 17)
Matching
1. d   2. e   3. a   4. b   5. c

Fill in the Blanks
6. minerals
7. tetrahedron
8. families
9. elements, aluminum
10. metamorphic, sedimentary

Multiple Choice
11. d   12. a   13. c

Knowledge Builder (page 18)
Activity #1:   1. sugar: hexagonal   2. salt: cubic

Inquiry Investigation (page 19–20)
Answers may vary but should include: Minerals can be identified according to their physical properties. Mineralogists commonly use a battery of tests to determine the identity of minerals: streak, hardness, luster, cleavage, specific gravity, magnetism, and reaction to acid. Streak—color of streak mineral leaves when dragged across a piece of white porcelain plate. Hardness—test for how hard a mineral is using the Mohs' scale. Luster—test for how light interacts with the surface of the mineral. Cleavage—a test to determine how a mineral breaks. Specific Gravity—tests the weight of an amount of mineral compared to an equal amount of water. Magnetism—test to see if mineral is attracted to a magnet. Reaction to acid—tests the reaction of mineral to vinegar (usually fizzing/bubbling).

Plate Tectonics
Quick Check (page 24)
Matching
1. e   2. d   3. a   4. b   5. c

Fill in the Blanks
6. Faulted mountains
7. Volcanic mountains
8. Continental Drift
9. Plate Tectonics, Continental Drift
10. lithosphere

Multiple Choice
11. c   12. a   13. b

Knowledge Builder (page 25)
Activity #2: Answers may vary but should include the following information:
1. Divergent boundary—move apart; most are in the ocean. They build undersea mountain ranges called mid-ocean ridges.
2. Convergent boundary—crash together, often forming mountains and producing earthquake and volcanic activity.
3. Transform boundary—two plates slide past each other, often causing earthquakes.

Soil, Weathering, and Erosion
Quick Check (page 29)
Matching
1. b   2. c   3. e   4. a   5. d

Fill in the Blanks
6. Soil   7. minerals   8. alkaline
9. pH   10. crust

Multiple Choice
11. b   12. a   13. d

Knowledge Builder (page 30)
Activity #2:

Observation: Answers may vary but should include: Roots of the bean seeds start to penetrate the plaster, causing the plaster to crack or scale off.

Conclusion: Answers may vary but should include: Roots of plants break rocks into smaller pieces by physical forces. This process is called mechanical weathering.

Sedimentary Rocks
Quick Check (page 33)
Matching
1. d   2. a   3. e   4. b   5. c

Fill in the Blanks
6. Cementation   7. compaction, cementation
8. minerals   9. weathering
10. fossils

Multiple Choice
11. a   12. b

Knowledge Builder (page 34)
Activity #1:
Top—soil
Middle—sand
Bottom—gravel
Activity #2:
Conclusion: Answers may vary but should include: The model formed in layers and cemented together with plaster of Paris is similar to how sedimentary rock forms when a layer of sediment becomes solid. During the process of weathering, small pieces, or particles, break away from the main rock. Water flowing over the earth's surface picks up sediment or small pieces of rock, sand, clay, and other materials. The water flows into streams or rivers. As the flow slows down, some particles of rock and other materials fall to the bottom of the river and settle out of the water. After thousands of years, layers of sediment become solid through the processes of compaction and cementation, forming rocks.

Igneous Rocks
Quick Check (page 38)
Matching
1. b   2. a   3. d   4. e   5. c
Fill in the Blank
6. Magma   7. rate, cool
8. crystal size, arrangement   9. minerals
10. plate boundaries
Multiple Choice
11. b   12. c   13. a
Knowledge Builder (page 39)
Activity #1:
Conclusion: Answers may vary but should include: An igneous rock is formed by the cooling of melted material such as magma inside the earth and lava above the ground.
Activity #2:
Conclusion: Answers may vary but should include: Magma rises and erupts through the vent. In some eruptions, large chunks of hot rock are blasted high into the air and fall back into the lava flow. Gradually, the lava piles up around the vent, forming a volcanic mountain.
Inquiry Investigation (page 40-42)
Conclusion: Answers may vary but should include viscosity is the resistance of a substance to flow. Fluids range in viscosity. Heating reduces viscosity. For example, warm peanut butter flows more easily than cold.
Infer: Heat affects the viscosity of magma. The hotter the magma, the more easily it flows.

Metamorphic Rocks
Quick Check (page 45)
Matching
1. c   2. e   3. a   4. d   5. b
Fill in the Blanks
6. Heat, pressure, fluids   7. Metamorphic rock
8. asthenosphere, crust   9. metamorphism
10. converging
Multiple Choice
11. d   12. a   13. d
Knowledge Builder (page 46)
Activity #1:
Conclusion: Answers may vary but should include: The s'more was made by pressure and heating. Metamorphic rock is the type of rock formed when sedimentary rock or igneous rock undergo a change due to pressure or heat in the earth.
Activity #2:
Conclusion: Answers may vary but should include: The books represent layers of rock building up on top of the rock particles (clay balls) causing heat to build as the rock particles are pushed deeper into the earth's crust. Metamorphic rock is the type of rock formed when sedimentary rock or igneous rock undergo a change due to pressure or heat in the earth.

The Rock Cycle
Quick Check (page 49)
Matching
1. d   2. e   3. a   4. b   5. c
Fill in the Blanks
6. lithosphere   7. Igneous rocks
8. metamorphosed   9. phases
10. asthenosphere
Multiple Choice
11. a   12. b   13. b
Knowledge Builder (page 50)
Activity #1:
Conclusion: Metamorphic

Sand
Quick Check (page 53)
Matching
1. c   2. e   3. a   4. b   5. d
Fill in the Blanks
6. water, wave   7. rocks, minerals
8. water   9. Spits
10. Beaches
Multiple Choice
11. d   12. a   13. b

Knowledge Builder (page 54)

Activity #1:

Conclusion: Sand consists of small rock and mineral particles ranging in size from 0.06 to 2 mm in diameter.

Activity #2:

Observation:

1. Sand builds up at the back of the box.
2. Sand builds up in front of the pencil.

Conclusion: Answers may vary but should include: Sand builds up at the back of the box and in front of the pencil because of the air from the fan. Mounds of wind-blown sand are called dunes. Sand dunes are simply piles of sand. Dunes are formed when wind and waves transport sand onto the beach.

Fossils

Quick Check (page 58)

Matching

1. d   2. c   3. e   4. a   5. b

Fill in the Blanks

6. amber   7. mammoths
8. Fossils   9. plants, animals
10. petrified

Multiple Choice

11. c   12. b   13. d

Knowledge Builder (page 59)

Activity #1:

Conclusion: Answers may vary but should include: The impressions left by plants or animals provide information about the size and shape of life on Earth in the past.

Activity #2:

Conclusion: Answers may vary but should include: The cast provides information about the size and shape of life on Earth in the past.

Bibliography

Student Literature Resources:

Barrow, Lloyd H. (1991). *Adventures with Rocks and Minerals.* Hillside, NJ: Enslow Publishers.

Beattie, Laura. (1996). *Discover Rocks and Minerals.* Mankato, MN: Creative Company.

Chesterman, Charles. (1979). *National Audubon Society Field Guide to North American Rocks and Minerals.* New York, NY: Alfred A. Knopf.

Christian, Spencer and Felix, Antonia. (1998). *Is There a Dinosaur in Your Backyard?* Hoboken, NJ: John Wiley & Sons, Inc.

Comfort, Iris Tracy. (1964). *Earth Treasures: Rocks and Minerals.* Englewood Cliffs, NJ: Prentice-Hall, Inc.

Downs, Sandra. (1999). *Earth's Hidden Treasures.* Brookfield, CT: 21st Century Books.

Fuller, Sue. (2002). *1001 Facts about Rocks and Minerals.* New York, NY: DK Pocketbooks, Inc.

Horenstein, Sidney. (1993). *Rocks Tell Stories.* Minneapolis, MN: Millbrook Press.

Jennings, Terry. (1995). *Rocks and Soil.* New York, NY: Children's Press.

Kittinger, Jo S. (1998). *A Look at Rocks from Coal to Kimberlite.* New York, NY: Franklin Watts, Inc.

Lambert, David. (1986). *Rocks and Minerals.* New York, NY: Franklin Watts, Inc.

Olifee, Neesha, Olifee, Jon, and Raham, Gary. (2000). *The Deep Time Diaries.* Golden, CO: Fulcrum Publishing.

Parker, Steve. (1997). *Rocks and Minerals.* New York, NY: DK Children, Inc.

Pellant, Chris. (2000). *Rocks and Minerals.* New York, NY: DK Children, Inc.

Pough, Frederic H. (1998). *Rocks and Minerals.* Boston, MA: Houghton Mifflin Company.

Rhodes, Frank H. T. (2001). *Geology.* New York, NY: Golden Press.

Ritter, Rhoda. (1977). *Rocks and Fossils.* New York, NY: Franklin Watts, Inc.

Rydell, Wendy. (1984). *Discovering Fossils.* Mahwan, NJ: Troll Associates.

Symes, R. F. (2000). *Rocks and Minerals.* New York, NY: DK Children, Inc.

Symes, R. F. and Harding R. R. (2007). *Crystal and Gem.* New York, NY: DK Children, Inc.

Williams, Brian. (1993). *Mining.* Austin, TX: Raintree Steck-Vaughn Publishers.

Bibliography (cont.)

Curriculum Resources:

Illinois State Geological Survey, GeoActivities Series. Department of Natural Resources, Natural Resources Building, 615 East Peabody Dr., Champaign, IL 61820.

DSM II Earth Science: Rocks and Minerals, Grades 5–6. Delta Science Module (see also Erosion). http://www.delta-education.com

Science & Technology for Children: Rocks and Minerals, Grade 3. National Science Resources Center ISBN 0-89278-746-5 Carolina Biological Supply, 2700 York Road, Burlington, NC 27215. http://www.carolina.com/

Plate Tectonics, The Way the Earth Works, Grades 6–8. Great Explorations in Mathematics and Science (GEMS). ISBN 0-924886-60-9 Lawrence Hall of Science, University of California, Berkeley, CA 94720. http://www.lhs.berkeley.edu/gems/

Stories in Stone, Grades 4–9. Great Explorations in Mathematics and Science (GEMS). ISBN 0-912511-93-1 Lawrence Hall of Science, University of California, Berkeley, CA 94720. http://www.lhs.berkeley.edu/gems/

Down to Earth, Grades 5–9. Activities Integrating Mathematics and Science (AIMS). ISBN 1-881431-00-2 AIMS Education Foundation, P.O. Box 8120, Fresno, CA 93747-8120. http://www.aimsedu.org/

Project Earth Science: Geology, Grades 5–8 by Brent A. Ford. ISBN 0-87355-131-1 National Science Teachers Association, 1840 Wilson Boulevard, Arlington, VA 22201-3000. www.nsta.org

Rocks & Minerals, Grades 7–12 by Doris Metcalf and Ron Marson. Task Oriented Physical Science (TOPS), 10970 S. Mulina Rd., Canby, OH 97013. http://www.topscience.org/

Classroom Resources:

Frey Scientific
905 Hickory Lane
P.O. Box 8101
Mansfield, OH 44901-8101
800-225-FREY (3739)
www.freyscientific.com

Ward's Natural Science
5100 W. Henrietta Rd.
P.O. Box 92912
Rochester, NY 14692-9012
www.wardsci.com

Bibliography (cont.)

Web Resources:

http://www.usgs.gov/ (highly recommended; see education link)

http://www.mii.org/

http://www.rocksforkids.com/

http://www.nearctica.com/

http://www.fi.edu/fellows/payton/rocks/index2.html

http://www.RockhoundingAR.com/pebblepups.html

State Geological Survey: Conduct a web search for your state geological survey, e.g., "Iowa State Geological Survey."

References:

Chesterman, C. (1979). *National Audubon Society Field Guide to North American Rocks and Minerals.* New York, NY: Alfred A. Knopf.

Christian, S., and Felix, A. (1998). *Is There a Dinosaur in Your Backyard?* New York, NY: John Wiley & Sons, Inc.

Cvancara, A. (1985). *A Field Manual for the Amateur Geologist.* New York, NY: John Wiley & Sons, Inc.

Dixon, D. and Bernor, R. (1992). *The Practical Geologist.* New York, NY: Simon & Schuster/Fireside.

Ford, B. (1996). *Project Earth Science: Geology.* Arlington, VA: National Science Teachers Association.

Fuller, S. (1995). *Rocks and Minerals.* New York, NY: Dorling Kindersley, Inc.

National Research Council. (1996). *National Science Education Standards.* Washington, D.C.: National Academy Press.

Pellant, C. (1992). *Rocks and Minerals.* New York, NY: Dorling Kindersley, Inc.

Pough, F. (1953). *A Field Guide to Rocks and Minerals.* Boston, MA: Houghton Mifflin Company.

Sager, R., Ramsey, W., Phillips, C., and Watenpaugh, F. (2002). *Modern Earth Science.* Austin, TX: Holt, Rinehart & Winston.

Tarbuck, E., and Lutgens, F. (1997). *Earth Science.* Upper Saddle River, NJ: Prentice-Hall.